PENGUIN BOOKS

THE TEN EQUATIONS THAT RULE THE WORLD

David Sumpter is Professor of Applied Mathematics at the University of Uppsala, Sweden. He is the author of *Soccermatics* and *Outnumbered*, which have been translated into ten languages, and *Collective Animal Behaviour*, the leading text in the academic field he helped create. He has worked with a number of the world's biggest football clubs, advising on analytics, as well as consulting on betting.

T0187301

DAVID SUMPTER

The Ten Equations that Rule the World

And How You Can Use Them Too

PENGUIN BOOKS

PENGUIN BOOKS

UK | USA | Canada | Ireland | Australia
India | New Zealand | South Africa

Penguin Books is part of the Penguin Random House group of companies
whose addresses can be found at global.penguinrandomhouse.com.

First published in Great Britain by Allen Lane 2020
Published in Penguin Books 2020

006

Printed and bound in Great Britain by Clays Ltd, Elcograf S.p.A.

The authorized representative in the EEA is Penguin Random House Ireland,
Morrison Chambers, 32 Nassau Street, Dublin D02 YH68

A CIP catalogue record for this book is available from the British Library

ISBN: 978-0-141-99109-2

Contents

Contents

List of Figures

Introduction: TEN

Is there a secret formula for becoming rich? Or for happiness? Or for becoming popular? Or for self-confidence and good judgement?

If you are browsing in your local bookshop and have picked up this book, or if you have just clicked on the 'Look inside' button in an online bookstore, then you will be aware that this is just one of many titles that offer you a formula for success in life.

Marie Kondo telling you to tidy up. Sheryl Sandberg telling you to lean in. Jordan Peterson telling you to stand up straight. Brené Brown encouraging you to be vulnerable. You are told you should calm the f*ck down, stop doing that sh!t, not give a f*ck, not be a miserable f*ck and make the most of your holy sh!t moment. You should get up early, make your bed, clear the path, do less, memorize more, declutter your mind, get things done, maximize your willpower, 'solve for happy', 'act like a lady and think like a man'. There are formulas for love, a science behind getting rich, a blueprint for success and five (or eight or twelve) rules for self-confidence. There is even, apparently, a miracle equation that claims to 'make impossible goals inevitable'.

All this advice presents a paradox. If it is all so simple, if there are seemingly easy formulas for getting everything we want from life, then why are all these books and lifestyle magazines full of tips with often contradictory messages? Why are there so many inspirational TV shows and TED talks offering motivational monologues? Why not just state the equations, give a few examples of how they work, and close down the self-help and smart-thinking industry? If it is all so mathematical, all so axiomatic, why not just tell us the answer – now?

As the number of suggested solutions to life's dilemmas increases, it becomes more and more difficult to believe that there is just one, or even a small number, of formulas for success. Maybe there really is no simple remedy to all the problems that life throws at us?

I want you to think about another possibility, one that this book explores. I am going to tell you a story about an exclusive society of individuals who have cracked the code. They have discovered a small number of equations – ten of them in fact – that can bring them success, popularity, wealth, self-confidence and good judgement. It is they who hold the secret, while everyone else continues to search for the answers.

This secret society has been with us for centuries. Its members have passed their knowledge down through the generations. They have taken positions of power in public service, in finance, in academia and, most recently, inside tech companies. They live among us, invisibly but powerfully advising us, and sometimes controlling us. They are rich, happy and self-assured. They have discovered the secrets that the rest of us so desire.

In Dan Brown's book *The Da Vinci Code*, cryptographer Sophie Neveu discovers a mathematical code while investigating the murder of her grandfather. She is led to Professor Robert Langdon, who reveals that her grandfather was head of a secret society, the Priory of Sion, which understands the world through a single number, the Golden Ratio, $\phi \approx 1.618$.

The Da Vinci Code is fiction, but the secret society I have investigated for this book has many similarities to the one described by Brown. Its secrets are written in a code that few fully comprehend and its members communicate in arcane scripts. The society has its roots in Christianity, and it has been torn by internal moral battles and conflicts. But it also, as we will soon find out, differs in important ways from the Priory of Sion. Unlike that society, it has no rituals whatsoever. This makes it far more difficult to detect and much more pervasive in its activities. It is invisible to those on the outside.

How, then, do I know about it? The answer to that is simple. I'm a member. I have been involved in its workings for twenty years and have gravitated closer and closer to its inner circle. I have studied the society's works and put its equations into practice. I have experienced at first hand the success that access to its code can bring. I have worked at the world's leading universities and was appointed full professor in applied mathematics the day before my thirty-third birthday. I have solved scientific problems in fields ranging from ecology and biology to political science and sociology. I have been a consultant to those inside government, finance, artificial intelligence, sports and gambling. And I'm happy – partly as a result of my success, but mainly, I believe, because of how the secrets I have learnt have shaped my own thinking. The equations have made me a better person: more balanced in my outlook and better able to understand the actions of others.

Membership of this club has brought me into contact with others like me. People like Marius and Jan, young professional gamblers who have found an edge on the Asian betting markets; those like Mark whose micro-second calculations skim profits from small inefficiencies in share prices. I have worked with the data scientists at the football club Barcelona who study how Lionel Messi and company control the field of play. I have met the technical experts employed by Google, Facebook, Snapchat and Cambridge Analytica, who control our social media and are building our future artificial intelligence. I have witnessed at first hand how researchers like Moa Bursell, Nicole Nisbett and Viktoria Spaiser use equations to detect discrimination, understand our political debates and make the world a better place. I have learnt from the older generation, like ninety-four-year-old Oxford professor Sir David Cox, who have discovered the code on which the secret society is built.

Now I am ready to name the secret society to which I and they belong. I will refer to it as 'TEN', based on the number of equations a fully qualified member needs to know. I am ready to reveal its secrets – to tell you the Ten Equations.

The problems TEN addresses include everyday dilemmas. Should you quit your job (or your relationship) and try something else? Why do you feel that you are less popular than those around you? How much effort should you put into becoming more popular? How can you best cope with the vast flood of information from social media? Should you let your kids spend six hours a day staring at their phones? How many episodes of a Netflix series should you watch before trying something else?

These might not be the problems you expect a secret society to be resolving. But here's the thing. The same small set of formulas can provide the answers to questions ranging from the trivial to the profound; and about you as an individual and society as a whole. The confidence equation, introduced in Chapter 3, which helps you decide whether you should quit your job, also allows professional gamblers to know when they have an edge on the betting markets and reveals subtle racial and gender biases at work. The reward equation, discussed in Chapter 8, shows how social media has driven society to tipping point and why this isn't necessarily a bad thing. By understanding how this and other equations are used by Internet giants to reward us, to influence us and to classify us, we can better balance our own and our children's use of social media, games and advertising.

We know these equations are important because of the success they have brought the people who already use them. Chapter 9 tells the story of three engineers in California who used the learning equation to increase the time viewers spent watching YouTube by 2,000%. The betting equation, the influencer equation, the market equation and the advertising equation have reshaped, respectively, betting, technology, finance and advertising to generate billions of dollars of profits for a small number of TEN's members.

As you learn the equations in this book, more and more aspects of the world will start to make sense. When you see through the eyes of TEN, big problems become small and small problems become trivial.

If you are just looking for quick fixes, then there is, of course, a catch. To be admitted to TEN you will need to learn a new way of thinking. TEN asks you to break down the world into three categories: *data*, *model* and *nonsense*.

One of the reasons TEN is so powerful today is that we have more *data* than ever: movements of the stock exchange and betting markets; and personal data about what we like, buy and do collected by Facebook and Instagram. Government agencies have data on where we live, how we work, where our children go to school and how much we earn. Pollsters collect and synthesize our political views and attitudes. News and opinions are collated on Twitter, in blogs and news websites. Every movement of sports stars on the field of play is logged and stored.

This explosion of data is obvious to everyone, but TEN's members have recognized the importance of identifying mathematical *models* to explain the data. Like them, you can learn how to build models, to use the equations to take control of and use data in a way that gives you an edge, a small advantage, over others.

The final category, *nonsense*, is something we need to learn to spot. You will come to understand that, as enjoyable and fulfilling as it can be to talk nonsense, and while we all do it a lot of the time, you will need to put it aside when you think like a member of TEN. We need to call out nonsense whenever we encounter it, no matter who voices it. I will show you how to ignore nonsense, and refocus on data and model.

This is not just a self-help book. It is not the Ten Commandments. It is not a list of dos and don'ts. There are equations in this book, but no recipes. You can't simply skip to page 157 and find out the exact number of Netflix episodes you need to watch.

Rules and recipes exploit our fears. Instead of building on those fears, this book explains how the code of TEN has evolved over and shaped the last 250 years of human history. We are going to learn from the mathematicians who developed the code and understand the philosophy that underlies their thinking. Learning TEN challenges many of our everyday assumptions. It involves rethinking terms such as 'political correctness', re-evaluating the judgements we make about others and reconsidering the stereotypes we create.

This is also a tale about morality, because it would be wrong of me to reveal so many secrets without investigating the effect the society of TEN has had on the world. If a small group of people can steer the rest of us, then we need to know what motivates the choices they have made. The story I tell here has forced me to re-evaluate myself and what I do. It has forced me to ask myself whether TEN is good or evil, and to think about the moral rules we should set up for ourselves in the future.

When handing down his power to a new generation, Spider-Man's uncle tells him that 'with great power comes great responsibility'. With so much at stake, the hidden powers of TEN require even greater responsibility than those bestowed by a Spider-Man suit. You are about to learn secrets that can transform your life. And you will also be forced to think about the effect these secrets have had on the world we live in.

For too long, the code has only been accessible to a chosen few. Now we are going to talk about it, openly and together.

I

The Betting Equation

$$P(\text{favourite wins}) = \frac{1}{1 + \alpha x^\beta}$$

The thing that first struck me about Jan and Marius as we shook hands in a hotel lobby was that they were not much older than the students I teach at university. And here I was, hoping to learn as much from them about the world of gambling as they presumably hoped to learn from me about maths.

We had chatted online, but this was the first time we had met in person. They had flown in to see me as part of a kind of European tour, meeting football betting experts and tipsters one by one to prepare their own strategy for the coming year. My hometown of Uppsala, Sweden, was their final stop.

'Shall we take our laptops with us to the pub?' Marius asked me as we prepared to leave the hotel.

'Of course,' I replied.

This may well have been a 'getting to know each other' meeting, before we started work properly the next day, but all three of us knew that even the most informal of discussions would require some number-crunching. The laptops always had to be on standby.

You might think that successful football gambling requires a lot of knowledge; that you need to have an in-depth understanding of the game, including knowing the form of each of the players, having some insight into injuries and perhaps getting your hands on some insider information. A decade ago, that might well have been the case. At that time, carefully watching the matches, observing the players' body language and seeing how they performed in one-to-one situations might have given you an edge over the punters who naively backed the home favourite. But not any more.

Jan had only a cursory interest in football and he had little intention of watching the majority of matches we were going to bet on in the upcoming 2018 World Cup. 'I'll enjoy the Germany matches,' he said with a confident smile.

It was the evening of the opening ceremony, the start of an event that few people on the planet, with or without an interest in football, could avoid hearing about. But aside from Jan's interest in his own national team, to him it was all the same – Bundesliga, Norwegian Tippeligaen or the World Cup; tennis or horses. Each tournament and each sport was just another opportunity for him and Marius to make money. And it was their search for these opportunities that had brought them to me.

A few months earlier I had published an article about my own football betting model.[1] This was no ordinary mathematical model. At the start of the 2015–16 Premier League season, I wrote down a single equation and proposed that it could beat the bookmakers' odds for Premier League match outcomes. It did.

By May 2018 it had racked up a 1,900% profit. If you had invested £100 in my model in August 2015, then less than three years later you would have £2,000. All you would have had to do was unthinkingly place the bets my model suggested.

My equation had nothing to do with what happened on the pitch. It certainly didn't involve watching the matches and it definitely didn't care who won the World Cup. It was a mathematical function that took in the bookmaker's odds, adjusted them slightly for a historical bias, and suggested new odds on which to bet. That was all that was needed to win money.

I had been completely open about my equation and it had garnered quite a bit of attention. I had published the details in *The Economist*'s *1843* lifestyle magazine and had talked about it in interviews with the BBC, CNBC, with newspapers and on social media. It was hardly a secret. It was this model that Jan and Marius were asking me about now.

'Why do you think you still have an edge?' asked Marius.

The currency of gambling is information. If you know something that other people don't know and that piece of information makes money, then the last thing you want to do is share it. The term 'edge' refers to that small informational advantage you have over the book-makers. In order to protect your edge, you should keep it a secret. If the money-making scheme gets out, then others will exploit it and the bookmakers will correct their odds. Your edge will disappear. That is the theory, anyway. But I had done the opposite. I had done every-thing possible to tell people about my equation. Marius was wondering why, despite the publicity, my model still worked.

A large part of the answer to Marius's question can be found by scrolling through the emails and DMs I get every day asking for bet-ting tips: 'Who do you think is going to win tomorrow's match? Read a lot about you and started believing in you'; 'I intend to raise funds to help me build capital to start a business. Your football tips would definitely lead me in the right direction'; 'Who have you got – Croatia or Denmark? I have a gut feeling Denmark will pull it off, but I'm not too sure'; 'What do u think the result gonna be in the England match? draw?' The requests go on and on.

I don't feel particularly good about saying this, but the reason that people keep sending me these messages also answers Marius's ques-tion. Despite my efforts to outline the limitations of my approach and my emphasis on a long-term strategy based on statistics, the public's response consisted largely of messages asking things like 'Will Arsenal win at the weekend?' or 'Will Egypt qualify from the group-stage if Salah doesn't play?'

It gets worse. The people who email me have at least searched the Internet for maths and gambling advice. There are many more who gamble without doing any research. They are gambling on a gut feeling, gambling for fun, gambling because they are drunk, gam-bling because they need cash and (in some desperate cases) gambling because they can't stop themselves. And, in total, there are many more of them than the small group of informed gamblers who are using my method or something similar.

'The reason the model continues to win is that it suggests bets people don't want to make,' I explained to Marius. 'Betting on a draw when Liverpool play at Chelsea or backing Manchester City to beat Huddersfield at small odds isn't fun.' It takes time and patience to make a profit.

Marius's first email to me fell into the 1% of messages that are different. He told me about an automated system that he and Jan had developed to find value in betting markets. Their idea was to exploit the fact that the majority of bookmakers are 'soft', that they offer odds that don't always reflect the true probability of a team winning.

The vast majority of punters (most likely including all those who were sending me messages asking for match tips) use 'soft' bookmakers. High street names like Paddy Power, Ladbrokes and William Hill are soft, as are smaller online names, such as RedBet and 888sport. These bookmakers prioritize special offers to encourage customers to gamble, but pay less attention to creating odds that reflect the probable outcome of sporting events. This latter activity, of accurately tuning odds to predict match outcomes, is performed by 'sharp' bookmakers such as Pinnacle or Matchbook, which are typically used by the other 1% of gamblers.

Marius and Jan's idea was to use the 'sharp' bookmakers to cream money from the 'soft' bookmakers. Their system monitored the odds at all bookmakers, soft and sharp, and looked for discrepancies. If one of the soft bookmakers was offering more generous odds than the sharp ones, then their system would suggest a bet at that particular soft bookmaker. This strategy by no means guaranteed a win, but since sharp bookmakers were more accurate, it gave Jan and Marius their all-important edge. In the long term, over hundreds of bets, they would win money at the soft bookmakers.

There was one limitation to Jan and Marius's system: the 'soft' bookmakers ban winners. It is the bookmakers who decide if they want your custom, and as soon as they saw that Jan and Marius's accounts were making a profit, they would ban them. 'You are now restricted to a maximum bet size of £2.50,' the message from the bookmakers would read.

But the guys had found a way to get their own back. Having developed their system, they were now offering a subscription service. For a monthly fee, subscribers were alerted via a direct link to value-for-money bets at soft bookmakers. This meant that Jan and Marius could continue to profit even if they themselves got banned. It was win-win for all involved, apart from the bookmakers. Part-time gamblers could get tips that would win in the long term, and Jan and Marius took a cut.

This was why I was sitting here in the pub with the two of them. They had mastered the art of gathering data and placing bets automatically. I had developed an equation that could further improve their edge: my model of the Premier League could not only beat the soft bookmakers, but also the sharp ones.

At this point I believed that I'd found an edge in the upcoming World Cup. But I needed more data to test my hypothesis. Before I finished telling them about my idea, Jan had his laptop open and was trying to get on the pub Wi-Fi. 'I'm sure I can get the qualifying odds and the odds from the last eight major international tournaments,' he said, 'I've got some code that will scrape [the term for automatically scanning web pages and downloading data] them for us.'

By the time we'd finished our drinks we had a plan and had identified the data we needed to carry it out. Jan went back to his hotel and set up his computer to scrape historical odds overnight.

*

Both Jan and Marius are of a new breed of professional gambler. They can programme a computer, they know how to get hold of data and they understand maths. Their type is often less interested in a particular sport and more interested in the numbers than the old-school gamblers. But they are just as interested in making money and are much better at it.

The betting edges which I had uncovered had put me on the pair's radar and let me into the peripheries of their gambling network. But from the cautious answers they gave at the pub when I asked about other projects they were working on, I could tell that full membership to their club wasn't on offer. Not yet, anyway. I was an amateur – they had laughed at me when I said I planned to place £50 bets on the system we were developing – and information about their other projects was shared on a need-to-know basis.

I had another contact, though, who had already been more open with me. He had recently left the sports trading industry, and while he didn't want me to reveal his identity and who he had worked for – we'll call him James – he was happy to share his experiences.

'If you have got a genuine edge, then the only limit to how fast you can make money is how quickly you can place bets,' James told me.

To understand James's point, first imagine a traditional investment with a 3% rate of return. If you invest a total capital of £1,000, then one year later you will have £1,030, a profit of £30.

Now consider gambling with £1,000 and an edge of 3% over the bookmakers. You certainly don't want to risk all your capital on one bet. So, let's start with a bet of £10, a relatively modest risk. You won't win every bet, but a 3% edge means that, on average, you will win 30p per bet on a £10 stake. The rate of return on your £1,000 investment is thus 0.03% per bet.

To expect to make a profit of £30 you will need to place one hundred £10 bets. One hundred bets per year, or roughly two per week, is more than most of us would place. For us amateurs, it is sobering to understand that, even if you do have an edge, as a casual gambler you can't expect to make much from a £1,000 capital investment.

The guys that James worked with weren't casual gamblers. Across the world, there are easily 100 football matches per day. Jan had recently downloaded data for 1,085 different leagues. Add to that tennis, rugby, horses and every other sport under the sun and there are a lot of betting opportunities out there.

Let's imagine for now that James and his colleagues just have an edge on football, and bet on those 100 matches per day throughout the year. Let's also assume, as the profits roll in, that they increase their stake size in proportion to their bankroll, so once they have made £10,000 they stake £100 on each bet. At £100,000 the stakes are £1,000 and so on. How much will gamblers with a 3% edge have made by the end of the year? £1,300, £3,000, £13,000 or £310,000?

By the end of the year they should have £56,860,593.80. Nearly £57 million! Each bet multiplies the capital by only 1.0003, but after 36,500 bets the power of exponential growth kicks in and the profits multiply dramatically.[2]

In practice, however, this level of growth isn't achievable. Even if the sharp bookmakers used by James and his former colleagues allow larger bets than soft bookmakers, there are still limits. 'The betting companies in London have grown so quickly and become so huge that they have to place their bets through brokers now. Otherwise, if it becomes widely known that they are placing a certain bet on a certain match, then everyone else floods into the market and their edge disappears,' James told me.

Despite these limitations, the money is still flowing into equation-driven betting companies. You don't need to look any further than the stylish interiors of their London offices to see the evidence of their success. The employees of one of the industry leaders, Football Radar, start the day with a free breakfast, enjoy access to a luxury gym, can take a break to play table tennis or PlayStation and are provided with all the computer equipment they need and want. The data scientists and software developers are encouraged to work their own hours and the company claims to provide the type of creative environment usually associated with Google or Facebook.

Football Radar's two big competitors, Smartodds and Starlizard, are also based in London. These companies are owned by, respectively, Matthew Benham and Tony Bloom, whose careers have progressed through their skill with numbers. Benham studied at Oxford, where he started his statistics-based gambling operation, while Bloom has a background as a professional poker player. In 2009 both of them bought the football clubs of their home towns, with Bloom buying Brighton & Hove Albion and Benham buying Brentford FC. Once Benham had established himself as consistently ahead of the game, he decided it was best to own the game too, and added sharp bookmakers, Matchbook, to his assets.

Benham and Bloom both found small edges using big data and made massive profits.

*

The edge that I proposed to Jan and Marius for the probability that the favourite wins a World Cup match is based on the following equation:

$$P(\text{favourite wins}) = \frac{1}{1 + \alpha x^\beta}$$

(Equation 1)

where x is the bookmaker's odds of the favourite winning. Here the odds are given in UK format, so that odds of 3 to 2 or $x = 3/2$ means that for every £2 staked £3 will be paid out if the bet is won.

Let's break down what Equation 1 actually says. We start with the left-hand side where I have written 'P(favourite wins)'. A mathematical model never predicts 'win' or 'lose' with absolute certainty. Instead, 'P(favourite wins)', the probability that the favourite wins, is a value between 0% and 100% predicting the level of certainty I assign to the outcome.

This probability depends on what we put into the right-hand side of the equation, which contains three letters, x from the Latin alphabet, α and β from the Greek. A student once told me that she thought that maths was straightforward when it came to dealing with xs and ys in the Latin alphabet, but that it got difficult when we started talking about αs and βs in the Greek. To mathematicians this is funny because x, α and β are just symbols, they don't make the maths more or less difficult, and I think the student was joking. But at the same time she made an important point: when α and β appear in equations the maths itself tends to be more difficult.

So, let's start without them. The equation

$$P(\text{favourite wins}) = \frac{1}{1 + x}$$

is much easier to understand. If, for example, the odds were 3/2 (2.5 in European odds, +150 in US odds) then the probability that the favourite wins is

$$P(\text{favourite wins}) = \frac{1}{1 + \frac{3}{2}} = \frac{2}{2 + 3} = \frac{2}{5}$$

In fact, this equation, without the α and β, tells us the bookmaker's estimate of a win for the favourite. They believe that the favourite has a 2/5 or 40% chance of winning the match. In the other 60% of cases, the match will end either in a draw or with the underdog winning.

Without the α and β (or strictly speaking with $\alpha = 1$ and $\beta = 1$) my betting equation is relatively straightforward to understand. But without the α and β this equation will not make any money. To see why, think about what would happen if you bet £1 on the favourite. If the bookmaker's odds are correct, then you will win £1.50 two times out of five and you will lose one dollar three times out of five. On average you can expect to win

$$\frac{2}{5} \times \frac{3}{2} + \frac{3}{5} \times -1 = \frac{3}{5} - \frac{3}{5} = 0$$

In words, the equation tells you that you should expect, on average, after lots of betting, to win nothing. Zero. Zilch. Except it is much worse than that. At the start I assumed the odds offered by the book-makers were fair.[3] In reality, they aren't fair at all. The bookmakers always adjust their odds in order to tip things in their favour. So instead of offering 3/2, they will offer 7/5, for example. It is this adjustment that ensures that, unless you know what you are doing, they will always win and you will always lose. With 7/5 odds you will lose, on average, 4 cents per bet.[4]

The only way to beat the bookmakers is to look at the numbers, and this data was exactly what Jan's computer had spent the evening scraping together after we had left the pub. It had collected odds and results for every World Cup and Euros match, including qualifiers, since the tournament in Germany in 2006. In the morning, sitting in my office at the university, we started looking for an edge.

We first loaded the data and looked at it inside a spreadsheet like the one below.

Favourite	Underdog	Odds for favourite wins (x)	Bookmakers' probability favourite wins $\frac{1}{1 + x}$	Favourite won? ('yes' = 1, 'no' = 0)
Spain	Australia	11/30	73%	1 (win)
England	Uruguay	19/20	51%	0 (lost)

Switzerland	Honduras	13/25	66%	1 (win)
Italy	Costa Rica	3/5	63%	0 (lost)
...				

From these historical results, we can get an idea of how accurate the odds are by comparing the last two columns of the spreadsheet data above. For example, in the match between Spain and Australia in the 2014 World Cup, the odds predicted a 73% probability that Spain would win, which they did. This could be considered a 'good' prediction. On the other hand, before Costa Rica beat Italy, the odds gave the Italians a 63% probability of winning. This could be considered a 'bad' prediction.

I use the words 'good' and 'bad' in quotation marks here because we can't really say how good or bad a prediction is without having an alternative to compare it against. This is where α and β come in. These are known as the parameters in Equation 1. Parameters are the values that we can adjust in order to fine-tune our equation, to make it more accurate. While we cannot change the closing odds for the Spain v. Australia match, and we certainly can't influence the result of the match between the two nations, we can choose α and β in order to make better predictions than the bookmakers.

The method of searching for the best parameters is known as logistic regression. To picture how logistic regression works, first consider how we could improve our predictions of the Spain v. Australia match by adjusting β. If I make $\beta = 1.2$, and keep $\alpha = 1$, then I get

$$\frac{1}{1 + \alpha x^\beta} = \frac{1}{1 + (\frac{11}{30})^{1.2}} = 77\%$$

Since the outcome was a win for Spain, this prediction of 77% can be considered better than the 73% predicted by the bookmakers.

There is a problem, though. If I increase β, then I also increase the predicted probability for an England win against Uruguay from 51% to 52%. And England weren't that lucky against Uruguay in 2014. To address this problem, I can increase the other parameter, setting $\alpha = 1.1$, while keeping $\beta = 1.2$. Now, the equation predicts that the probability of Spain beating Australia is 75% and the probability of England beating Uruguay is 49%. Both of the match predictions have improved from when we set α and β both equal to one.

I have just looked at one adjustment to each of the parameters α and β and compared the result to only two matches. Jan's data set consisted of the 284 matches at every World Cup and Euros since 2006. For a human, it would be very time consuming to repeatedly update parameter values, put them into the equation and see whether they improve our predictions or not. We can, however, use a computer algorithm to perform this calculation, and this is what logistic regression does (see Figure 1). It systematically adjusts the values of α and β to give predictions that lie as close as possible to the actual match outcomes.

I wrote a script in the programming language Python to perform the calculations. I pressed 'run' and watched my code crunch through the numbers. A few seconds later I had a result: the most accurate predictions were made when $\alpha = 1.16$ and $\beta = 1.25$.

These values caught my attention immediately. The fact that both parameters, $\alpha = 1.16$ and $\beta = 1.25$, were bigger than 1 indicated a complicated relationship between the odds and the results. The best way to understand this relationship is to add a new column to our spreadsheet, comparing our logistic regression model with the bookmaker predictions.

Favourite	Underdog	Odds for favourite wins (x)	Bookmakers' probability favourite wins $\frac{1}{1 + x}$	Logistic regression probability favourite wins $\frac{1}{1 + 1.16x^{1.25}}$	Favourite won? ('yes'= 1, 'no'= 0)
Spain	Australia	11/30	73%	75%	1 (win)
England	Uruguay	19/20	51%	48%	0 (lost)
Switzerland	Honduras	13/25	66%	66%	1 (win)
Italy	Costa Rica	3/5	63%	62%	0 (lost)
. . .					

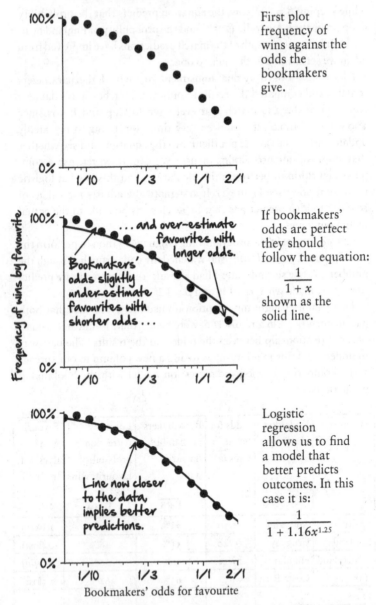

First plot frequency of wins against the odds the bookmakers give.

If bookmakers' odds are perfect they should follow the equation:
$$\frac{1}{1 + x}$$
shown as the solid line.

... and over-estimate favourites with longer odds.

Bookmakers' odds slightly under-estimate favourites with shorter odds ...

Logistic regression allows us to find a model that better predicts outcomes. In this case it is:
$$\frac{1}{1 + 1.16x^{1.25}}$$

Line now closer to the data, implies better predictions.

Bookmakers' odds for favourite

Figure 1: Illustration of how logistic regression estimates $\alpha = 1.16$ and $\beta = 1.25$

Here we can see that what hardened gamblers call a long-shot bias exists against strong favourites, like Spain. These teams were typically under-estimated by the bookmakers' odds and therefore worth backing. On the other hand, weaker favourites like England were over-estimated in 2014. The chance of England winning was lower than suggested by the odds. Although these differences between predictions and model were small, Jan, Marius and I knew that they were big enough for us to make a profit.

We had found a small edge on the World Cup. What we didn't know yet was whether the edge in previous tournaments would be present this time around. But we were prepared to risk a bit of money to find out. It took until lunchtime to implement a trading system based on my equation. We pressed 'Run' and put it into action. Our bets would be placed automatically throughout the World Cup.

After lunch, we went back to my house. Marius and I sat in my basement and watched Uruguay play Egypt. Jan took out his laptop and started to download tennis odds.

*

The betting equation isn't just about one World Cup, nor is it just about making money at the bookmakers. Its real power comes from the way it forces us to see the future in terms of probabilities and outcomes. Utilizing the betting equation means leaving our hunches behind and forgetting the idea that the outcome of a football match, a horse race, a financial investment, a job interview or even a romantic date can be predicted with 100% certainty: you can't know for sure what is going to happen.

Most of us have a vague idea that events in the future are, to a large degree, determined by chance. If a weather forecast tells you that there is a 75% chance it will be sunny tomorrow, then you shouldn't be too surprised if you find yourself caught in a downpour on your way to work. But finding the small edges hidden within probabilities requires a step up in your understanding.

If a particular outcome is important to you, then think about the probability that it will work out and the probability that it won't. I recently talked to a CEO of a very successful start-up that had grown through four rounds of multimillion-dollar investment and had one hundred employees, and he happily admitted that the chance of long-term profit for himself and his investors was still only around 1 in 10. He was working long hours and was fully committed, while simultaneously aware that everything could suddenly unravel.

When looking for the job of our dreams or the love of our life, success rates for individual applications or dates can be very low. There are always factors outside your control. I am often surprised by the way people who fail at the interview stage for a job beat themselves up about what *they* might have done wrong, rather than consider the fact that this was very likely the day when one of the other four applicants did everything right. Try to remember that 20% success probability you had before you walked into the building for your interview. Until you fail five or so interviews, there is little reason to get down about any particular outcome.[5]

Romance is trickier to quantify, but the same probabilistic principles apply. Don't expect Prince or Princess Charming to turn up on your first Tinder date, although do take some time to reflect over your approach when you are sitting alone after unsuccessful date number thirty-four.

Once you have identified the probabilities involved, think about how they relate to the size of the investment made and the potential profits. My advice to think probabilistically is not a call for karmic calmness or an attempt to get you to be more mindful. The CEO with a 1 in 10 chance of success had a business idea that had the potential to be the next Uber or the next Airbnb – an idea that could build a 10 billion-dollar company. Ten billion divided by ten is still one billion, a massive expected profit.

Thinking probabilistically is about being realistic in the face of odds that are often stacked against you. In horse-racing and football, long shots tend to be overvalued by naive gamblers, but in real life we tend to undervalue long shots. We are cautious and risk-averse by nature. Remember that the pay-off of getting a job you really enjoy or finding a partner you love is massive. This means that you need to be prepared to take big risks to achieve your goals.

*

Maths requires hard work and perseverance. Five minutes ago, I finished reading one of the most remarkable papers in the history of applied mathematics, an article that is literally worth $1 billion. And even though I knew the importance of the maths within the paper on first reading it, I found it more difficult when I got to the equations. I skipped over them, told myself I'd get the details later, and moved on to the interesting bits.

The article in question was 'Computer based horse race handicapping and wagering systems: a report' by William Benter.[6] It is a manifesto, a scientific statement of intent. And it is the work of a man obsessed with rigour and full of belief in what he is doing, a man committed to documenting his plan before he carries it out, to show the world that when he wins it won't be because of luck – it will be because of mathematical certainty.

In the late 1980s William Benter set out to beat the Hong Kong horse-racing market. Before he started his project, high-stakes gambling had been an activity for hustlers. These hustlers could be seen hanging around the Happy Valley and Sha Tin racetracks, as well as the Royal Hong Kong Jockey Club, trying to gather inside information from the owners, stable staff and trainers. They would find out whether or not the horse had eaten breakfast or been given extra workouts in secret. They would befriend jockeys and quiz them about their strategy for the upcoming races.

As an American, Benter was an outsider in this world, but he had identified another way of getting his hands on inside information, a way that the hustlers had missed, despite the fact it had been lying right there in the Jockey Club offices the whole time. Benter picked up copies of the track yearbooks and, with help from two secretaries, he started typing racing results into a computer. He then had what he later told *Bloomberg Businessweek* was his breakthrough moment. He took the closing odds, also collected by the Jockey Club, and digitized them too. It was these odds that allowed Benter to apply a method similar to the one I had shown to Jan and Marius: using the betting equation. This was the key to finding the inaccuracies in the gamblers' and the tipsters' predictions.

Benter didn't stop there. In the basic equation I presented in the last section, I was limited to identifying biases in the football odds.

Now, on my second or third read of his paper, I began to understand how Benter had been profitable over such a long period of time. In my own model, I didn't look at the additional factors that would allow me to predict match outcome. But Benter did go the extra mile for horse racing. His rapidly growing data set included past performance, time since last race, age of the horse, jockey's contribution, assigned post position, local weather and many other factors. Each of these factors was added term by term to the betting equation. As he included more and more detail, the accuracy of his logistic regression, and therefore his predictions, increased. After five person-years of data entry, his model was ready and, with capital raised by counting cards in casinos, Benter started to bet on the Happy Valley races.

During the first couple of months of gambling, Benter saw a 50% profit on his investment, but this profit disappeared again two months later. Over the next two years, Benter's profits bobbed up and down, sometimes nearing 100%, only to drop again to nearly zero. It was after about two and a half years that the model really started to pay off. The profits climbed to 200%, 300%, 400%, onwards and exponentially upwards. Benter told *Bloomberg Businessweek* that, in the 1990–91 season, he won $3 million.[7] The same publication estimated that over the next two decades Benter and a small number of competitors who used the same methods made over a billion dollars at the Hong Kong racetracks.

The most remarkable thing about Benter's scientific paper is not so much its contents, but the fact that very few people have read it. In the twenty-five years since its publication, it has been cited in other scientific articles a total of ninety-two times. To put that in context, an article I wrote fifteen years ago about how Temnothorax ants choose a new home has had 351 citations.

It is not just Benter's article that has been ignored. He references a paper written in 1986 by Ruth Bolton and Randall Chapman as 'required reading' for his own work.[8] Yet, nearly thirty-five years later, this inspirational article, showing how a profit could be made on American racetracks using the betting equation, has also been cited fewer than one hundred times.

Benter had no formal education in advanced mathematics, but he was prepared to put in the work required. He has been described elsewhere as a genius, but I don't see it that way. In my professional life, I meet and work with lots of non-mathematicians and non-geniuses who have persevered and learnt the same statistical methods Benter had used. They aren't usually gamblers. They are biologists, economists and social scientists who use statistics to test hypotheses. But they have taken the time to understand the maths.

I never get the maths on first reading. In fact, I have met very few professional mathematicians who can read and digest equations without having to go back through the details. And it is in those details that the secrets lie.

*

The biggest threat to any secret society is the threat of being uncovered. The modern version of the Illuminati conspiracy, which imagines tech-savvy overlords controlling the world's affairs, would require each and every one of the members to remain quiet about their goals and their methods. If just one of them shares the code or lets the society's plans slip out, then the whole enterprise is at risk.

This danger of discovery is the main reason most scientists don't give much credence to the existence of Illuminati-like organizations. Controlling all of human activity would require a large society and a massive secret. The risk of one of the members breaking down and telling all would be substantial.

But as we delve into the betting equation, we see how TEN's secret is hidden in plain sight. It is only as the society's members study and persevere that it slowly reveals itself to those who seek it out. The code is taught to everyone in schools and expanded on in courses at university, without us realizing what we are really learning. The society's members are only vaguely aware that they are part of this vast conspiracy. They feel they have nothing to reveal, nothing to confess and nothing to hide.

As a young prospective TEN member reads Benter's scientific article a second and a third time, she pushes herself to properly understand. She starts to feel a bond, a connection extending over decades and centuries. Benter surely felt the same connection as he studied the work of Ruth Bolton and Randall Chapman. In turn,

Bolton and Chapman before him would have had the same feeling as they studied the work of David Cox, whose proposal of logistic regression in 1958 laid the groundwork for them. Backwards in time, through Maurice Kendall and Ronald A. Fisher between the wars, all the way to the first ideas of probability expressed by Abraham de Moivre and Thomas Bayes in eighteenth-century London, the connection forged by mathematics stretches through history.

As she digs deeper into the details, our young acolyte sees that every part of the secret is there, revealed step by step on the pages in front of her. Benter has recorded the origins of his success in the code of equations. And now, twenty-five years later, she can pick apart that success, one algebraic symbol at a time.

It is the maths, the shared interest in the equation that pulls us together across great distances in time and space. Like Benter before her, our student is starting to learn the beauty of placing a bet based not on hunches, but on a statistical relation in data.

*

There is a way of explaining the idea behind the betting strategy Jan, Marius and I developed without using equations. In fact, I can explain the key idea in a single sentence: we found that the opening odds for the World Cup (those that the bookmakers give a long time in advance of matches) could be used to make better predictions of results than the closing odds (those that the bookmakers gave directly before the match).

This observation is counter-intuitive. When the bookmakers set their odds, there remains a lot of uncertainty about what will happen in the weeks (or months) before kick-off. Star players can be injured (as was Egypt's Mohamed Salah), a team might have a run of bad form (France drew with the USA a few weeks before the World Cup) or there might be a last-minute change of manager (as was the case for Spain). In theory, the odds should change to reflect these events, so that when Spain suddenly sacks their manager, their odds of beating Portugal drop.

The odds do change, but instead of changing to reflect the new reality, they tend to overshoot. As a match approaches, amateur gamblers enter the market, placing their bets and trying to predict match outcomes, and the bookmakers' odds change to reflect the bets these amateurs place. For example, the odds for France beating Peru lengthened from 2/5 to 1/2 in the build-up to their first group-stage match. Maybe some punters thought that if France couldn't beat the USA in a friendly, then Peru could sneak a point or even three. Other amateur gamblers undoubtedly read the newspaper criticism of mid-field star Paul Pogba and started to doubt his capacity to lead his national team to glory. Whatever the reasons, this was exactly the scenario which our model had found produced good value bets in previous World Cups. When the odds for strong favourites increased, then backing the favourite gave the edge. Our automated system detected the change in the odds, activated our betting function and placed £50 on France. After the match we had £75. It is as simple and as effective as that.

An important skill for the applied mathematician is to be able to explain the underlying logic behind the models we use. As Marius and I watched football during the afternoon after we had set up the model, we discussed why the odds became less accurate as we got closer to the World Cup.

'Most of our trading strategies are based on the idea that odds get more accurate closer to the match,' he told me. 'There must be something different about the World Cup.'

'It is the sheer volume of betting,' I speculated. 'There is a lot of football on TV, and it is fun to have a flutter. Some people bet out of national pride and others want to stick it to another nation.'

Marius agreed. The World Cup brought new audiences to football and they couldn't resist putting money on their personal favourite. We imagined loyal England fans thinking it would be fun to make some money at the expense of the French. We thought about Argentinians and Germans backing Switzerland against Brazil in their opener. As the money poured in on the underdogs, the bookmakers lengthened the odds for the favourites, and our model benefited by going against the crowd. Not every match would pay off for us – Brazil opened with a surprise draw against Switzerland – but history showed that backing very strong favourites directly before kick-off was most likely to make money.

The bias of amateur punters backing long shots was only part of our model. Our equation provided more subtle predictions: the values of $\alpha = 1.16$ and $\beta = 1.25$ meant that when there wasn't a very strong favourite, we should back the underdog instead, just like we'd seen when England lost to Uruguay in 2014. A good example of one of these predictions was the match between Colombia and Japan. The odds for Colombia to win lengthened from 7/10 to 8/9 in the days leading up to the match. Putting these odds into our equation suggested a bet on Japan was in order. This wasn't because Japan was more likely to win the match. It wasn't. Colombia remained favourites. Instead, the equation implied that Japan, who was now at odds of 26/5, was better value than Colombia. In this particular case, we got it right, Colombia lost and we won £260 from our £50 bet.

*

Sir David Cox is now ninety-five years old and has never stopped working. In a career spanning eight decades, he has authored 317 scientific articles and there are very likely more to come. From his office in Nuffield College, Oxford, he continues to write commentaries and reviews of modern statistics, as well as making new contributions to his field.

I asked him if he went into the office every day.

'Not every day,' he replied, 'not on Saturdays or Sundays.'

He then paused and corrected himself, 'I should say that the probability I come in on Saturday or Sunday is rather small. It can happen.'

Sir David Cox likes precision. His answers to my questions were careful and studied, and always qualified by stating the level of confidence he had in his ability to answer.

It was Cox who discovered the betting equation. Well, he would never say that, and it isn't completely accurate either. A more correct statement would be that he developed the theory of logistic regression, which I used to find values of α and β, and which Benter used to determine which factors predicted the outcome of horse races.[9] He developed the statistical method that allows the betting equation to make accurate predictions.

Logistic regression was a product of post-war Britain. Towards the end of the Second World War, as Sir David finished his education in mathematics at Cambridge University, he was seconded to work first for the Royal Air Force. Later he moved to the textile industry, as the UK began the process of rebuilding. He told me that his initial interest was in the abstract mathematics he had studied, but these placements opened his eyes to new challenges. 'The textiles industries were full of fascinating mathematical problems,' he said.

He admitted that his memory of specifics was vague, but his enthusiasm for those times shone through. He talked about how tests on various features of a material could be used to predict the probability it would break and the problems around creating a stronger, more uniform final product from roughly spun wool. These questions, combined with those he encountered at the Royal Air Force about the frequency of accidents and about wing aerodynamics, gave him plenty to think about.

It was from these practical questions that Sir David started to ask a more general question, a more mathematical question: what was the best way of predicting how an outcome – such as an aeroplane accident or whether or not a blanket would tear – was affected by various factors, like wind speed or the strains and stresses involved. This is the same type of question that Benter was asking about horses: what was the probability a horse would win as a function of its race history and the weather.

'The biggest controversies in the universities when I formulated the theory [in the mid-1950s] were about analysing medical and psychological data, predicting how different factors were related to a medical outcome,' Cox told me. 'Logistic regression came from a synthesis of my practical experience and my mathematical education. All the different problems I heard about from medicine, psychology and industry could be solved using the same family of mathematical functions.'

That family of mathematical functions turned out to be more important than even David Cox had imagined. From 1950s industry to its importance in interpreting the results of medical trials, logistic regression has been successfully applied to innumerable different problems. It is an approach now used by Facebook to decide which adverts to show us, by Spotify to recommend music to us, and as part of the pedestrian detection system in self-driving cars. And, of course, it is used in gambling . . .

I asked Sir David if he knew about Benter's success on the horses using logistic regression. He hadn't heard of him. So I told him about how logistic regression had made $1 billion. I then told him about Oxford student Matthew Benham and his success at predicting football match results.

'I prefer to say you should never gamble,' he told me after I had finished, before pausing and thinking for a very long time.

Then quietly he began to tell me a betting story of his own, about one of his colleagues in the 1950s, a story he asked me to promise never to repeat. And, I'm afraid to say, I'm going to keep to my word.

*

Betting isn't about predicting the future with certainty, it is about identifying small differences in the way you see the world and how others see it. If your vision is slightly sharper, if your parameters better explain the data, then you have an edge. Don't expect your edge to come straight away. It needs to be built up over time, through a process of trial and error, as you improve your parameter estimates. And don't expect to win all the time. In fact, only expect to win slightly more often than you lose as you play the game over and over again.

We sometimes tend to focus on our one 'Big Idea'. But what the betting equation tells us is that the key is to create different variations of our idea. Imagine you are starting your own yoga or dance class. Try different playlists with different groups and note down which gets the best response. By testing out lots of small ideas, we are running them off against each other like the horses at the Happy Valley racetrack. At the end of each race, we can reassess the winners and losers and look at the properties which led to success or failure.

If you are starting to test out a new idea, then you should perform what is known in the data science industry as A-B testing. When Netflix update their website design, they create two or more versions (A, B, C, etc.) and present them to different users. Then they look to see which design attracts the most engagement. This is a very direct application of the betting equation to the 'success' and 'failure' of design features. With all the traffic into Netflix, they can quickly build up a clear picture of what works and what doesn't.

You don't need to perform logistic regression in order to start making use of the betting equation. But, once you have understood the principle of tuning parameters to better fit data, progressing to learn the method itself is well within your grasp. Sir David Cox told me that he believes that most people could and should learn how to use the technique he developed. He told me that you don't need to understand all the mathematical details of how logistic regression works in order to understand what it reveals about the data you have collected.

*

I watched a lot of football during the World Cup, but, because I wasn't following the odds, I had no idea whether or not a result would earn me money. I just enjoyed the games. Now and again, Jan sent me automatically generated spreadsheets, with a list of bets placed and money won or lost. We lost during the first round of the group stage, but then started to win, and as the tournament went on the profits started to outweigh the losses. By the end of the World Cup, I'd made nearly £200 from a total of £1,400 worth of bets, a 14% rate of return on the investment.

After studying the updated spreadsheet containing our record, I looked again at the messages in my inbox, which had become more and more desperate as the World Cup progressed: 'Please, I know how you understand football tips and correct score so please can you help me?'; 'I wish to follow all your predictions to be able to help me bet correctly, as I have lost a whole lot to bookmakers'; 'There is a football jackpot in the country today with a lot of money. Helping me to win it, you will have helped a hundred people behind me.' The messages came almost hourly.

I couldn't help thinking about how our small profit came from their small pockets. The bookmakers were taking the biggest cut, of course, but the money Jan, Marius and I made was money that had previously belonged to someone else – maybe someone who didn't have very much to begin with.

It was then an idea started to build in my mind: the inequality between those who know the equations and those who don't isn't limited to gambling. Sir David Cox's statistical models have worked across many aspects of modern society. From the wool industry and aircraft design to modern data science and artificial intelligence, mathematical techniques have driven progress and been the basis of technology. This progress has been controlled by a very small proportion of the population: the ones who know the equations. And in many cases, it is the people who know the secrets who have benefited both socially and financially from their mathematics.

David Cox is a member of TEN. He doesn't know about his membership, but he invented one of the equations and he fully understands the other nine. As a result, his position in TEN's history is secured. He is a revered member of the highest order.

Benter, Benham and Bloom are part of TEN too. Maybe they don't know the equations in the formal, mathematical way that Cox understands them, but they understand the principles and they know how to put them into practice. Jan and Marius are on their way to joining them.

And me? I know the ten equations in the pure, unadulterated way in which academics understand them. I also know them in the street-smart way that Benter put them to use. And, although I hadn't acknowledged it before, I realized that TEN defines me, not just in how I work, but as who I am as a person.

2

The Judgement Equation

$$P(M|D) = \frac{P(D|M) \cdot P(M)}{P(D|M) \cdot P(M) + P(D|M^C) \cdot P(M^C)}$$

My friend Mark manages a team of financial traders, all with a background in maths or statistics. Mark has observed that his best traders have one thing in common: an ability to process and react to new information. As events occur they quickly adjust their understanding to the new reality.

The traders don't think in absolutes – such as 'this company is going to make a profit next quarter' or 'that start-up is going to fail' – they think in probabilities: 'there is a 34% probability the company will make a profit' or 'there is a 90% risk of failure for that start-up'. As new information comes in – for example, a CEO is forced to resign or the start-up's beta version has good traction – they update their probabilities: 34% becomes 21% and 90% becomes 80%.

I've heard similar stories from James, my contact in the gambling industry. They use variations of the betting equation, but with so much money on the line they also have to make snap decisions as to whether their model is valid for upcoming football matches. What do they do if one hour before the game the starting line-up changes and the assumptions underlying their model are no longer valid?

'It is at this point you find out who the really good traders are,' James told me. 'They don't overreact. One change in the starting line-up and the bet is still on; two to four changes and they start to weigh up the different possibilities; five or more changes and all bets are off.'

To learn to think like these analysts, first you have to put yourself in an emotionally stressful situation. When safely on the ground, most of us understand that flying isn't dangerous. When you step into a commercial plane, the probability you will die in a fatal crash is less than 1 in 10 million. But this can feel very different when you are up in the air.

Imagine you are an experienced traveller, having flown one hundred times before. But the flight you are now on is different. As you descend, the plane starts to rattle and shake in a way you have never experienced before. The woman next to you lets out a gasp; the man sitting across the aisle grips his knees. Everyone around you is visibly scared. Could this be it? Could the worst possible scenario be about to unfurl?

In situations like this, mathematicians will take a deep breath and collect up all the information they have. In mathematical notation we write the baseline probability of plane crashes as $P(\text{CRASH})$, with the P standing for probability and CRASH denoting the worst possible scenario of a fatal (for you) landing. We know from statistical records that $P(\text{CRASH}) = 1/10,000,000$, a 1 in 10 million chance.[1]

In order to understand how events depend on each other, we write $P(\text{SHAKE} \mid \text{CRASH})$ to denote the probability that the plane shakes this much *given* that it is just about to crash (SHAKE denotes 'the plane shaking', the vertical line, $\mid$, denotes 'given'). It is reasonable to assume that $P(\text{SHAKE} \mid \text{CRASH}) = 1$, that is, crashes are always preceded by a lot of shaking.

We also need to know $P(\text{SHAKE} \mid \text{not CRASH})$, the probability of this much shaking in an otherwise safe landing. Here you have to rely on your senses. This is the scariest flight you have been on, out of 100 similar trips, so $P(\text{SHAKE} \mid \text{not CRASH}) = 1/100$ is your best estimate.

These probabilities are useful, but they aren't what you are desperate to know. What you want to know is $P(\text{CRASH} \mid \text{SHAKE})$, the probability that you are about to crash given that the plane is shaking so much. This can be found using Bayes' theorem:

$P(\text{CRASH} \mid \text{SHAKE})$

$$= \frac{P(\text{SHAKE} \mid \text{CRASH}) \cdot P(\text{CRASH})}{P(\text{SHAKE} \mid \text{CRASH}) \cdot P(\text{CRASH}) + P(\text{SHAKE} \mid \text{not CRASH}) \cdot P(\text{not CRASH})}$$

The · symbol in the equation represents multiplication. Soon I will explain where this equation comes from, but for now let's just accept it. It was proved true by the Reverend Thomas Bayes in the mid-eighteenth century and has been used by mathematicians ever since. Putting all the numbers we have into our equation, we get:

$$P(\text{CRASH} \mid \text{SHAKE}) = \frac{1 \cdot \dfrac{1}{10000000}}{1 \cdot \dfrac{1}{10000000} + \dfrac{1}{100} \cdot \dfrac{9999999}{10000000}} \approx 0.00001$$

Even though this is the worst turbulence you have ever experienced, the chance you are going to die is 0.00001. There is a 99.999% chance that you are about to land safely.

The same reasoning can be applied in a whole range of different, seemingly dangerous situations. Even if you think you see something scary in the water while swimming on a beach in Australia, the probability that it is a shark remains minuscule. You may worry when a loved one is late coming home and you can't reach them, but the most likely explanation is that they have simply forgotten to charge their phone. A lot of what we consider as new information – shaking planes, murky shapes in the water or failed phone calls – is less scary than we think when we approach the problem correctly.

Bayes' rule allows you to properly assess the importance of information and to keep calm when everyone around you is panicking.

*

I see the world in a way I call 'cinema'. I spend a lot of my time, when I'm on my own or even when I'm in the company of others, playing films of my future in my head. These aren't just one film or one future, but they are movies of lots of different plot twists and endings. Let me try to explain using the aeroplane example.

When I am taking off and landing in an aeroplane, I see the crash I described above. If I am with my family in the plane, I imagine holding my children's hands, telling them I love them and not to worry. I imagine holding myself together for their sake as we plunge to our deaths. When I am alone on a plane, surrounded only by strangers, I see a different film. I see years stretching ahead, years of my family without me. The funeral passes by quickly and I see my wife alone with our kids, coping and telling them stories about our time together. This film is indescribably sad.

These films play, continually and in parallel, in an area of my brain just above my left eye. Or at least that is how it feels. Most of the movies are much less dramatic than the aeroplane crash. I have a meeting with a book editor and play through our discussion in my head, thinking about what I will tell her. I am delivering a seminar and I see how I will present the material, imagining the difficult questions that might come up. Many of the films are abstract: I navigate my way through a scientific article I am writing; I see the structure of the thesis of one of my PhD students; or I work through a maths problem. These types of films wouldn't work very well on the big screen. They are filled full of numbers, technical terms and scientific references. I enjoy them, but I'm very much a specialist audience.

I want to make sure that you don't get the impression that I view myself as an all-seeing oracle. That's not it at all. The films I create are fragmented. Details are missing, yet to be filled in by reality. Crucially, they are almost always wrong. The book editor takes the discussion in another direction and I forget my questions. A hole appears in the reasoning in a scientific article and I can't fix it. I make a massive calculation error in the first line of my maths and my results come out all wrong.

Psychologists have studied the ways in which people view the world and construct future narratives, but the scientific description of this process is not the point here. What is important is what *you* think about how *you* see the future. Is it in words, films or computer games? Is it in photographs, sounds or smells? Is it an abstract feeling or do you visualize real events? Try to identify the way you think about things. The way you see the world should remain personal – I don't want to change it. I would hate it if someone told me to switch off my movies. My 'cinema' is part of me.

Where mathematics comes into my thinking is in helping me to organize my film collection. The aeroplane crash is a good example. When I play the crash movie I also estimate the probability of its really happening and find it reassuringly low. This doesn't stop the film from playing. I still get scared when flying or when swimming in the ocean, but it does help me focus my thoughts. Instead of just being afraid, I think about how much my family means to me, and why I should travel less and swim in the ocean more.

The scientific term for the movies I play in my head is 'model'. The film crash is a model, a shark attack is a model and the plan for my scientific research is a model. Models can be anything from vaguely defined thoughts to more formally defined equations, like the one I developed for betting on the World Cup. The first step to a mathematical approach to the world is to become aware of how we use models.

*

Amy has started a new college course and is wondering who she should befriend and who she should stay clear of. She is quite a trusting person and the film she plays in her head is of other people welcoming her and being nice. But Amy is not completely naive. She knows by now that not everyone is nice and she has a 'bitch' film in her head too. Don't judge Amy on her choice of terminology, she keeps her thoughts in her own head after all. So, when she is introduced to Rachel, the girl sitting at the desk next to her, she assumes that the chance that Rachel is a bitch is pretty low, let's say, around 1 in 20.

I don't really think that Amy sets exact 'bitch' probabilities when she meets people. The reason I am setting a number now is to give a better feeling for the problem. You can think about it for a second and decide what proportion of the people you know are bitches. I hope it is less than 1 in 20, but feel free to choose your own numbers.

On that first morning, Rachel and Amy review some coursework from the lecture together. Amy is slow to grasp all of the details, as at her previous school they didn't cover the background to the concepts the lecturer is using. Rachel appears patient, but Amy can tell she is a bit frustrated. Why can't Amy learn faster? Then, just after lunch, a horrible thing happens. Amy is sitting in one of the toilet cubicles, browsing on her phone and minding her own business. She hears Rachel and another girl come in.

'That new girl is stupid,' Rachel says. 'I tried to explain "cultural appropriation" to her and she had no idea. She thought it was about white people learning to play the bongos!'

Amy sits very still, keeps quiet and waits for them to leave. What should she think?

Most of us would be sad or angry or both if we were in Amy's shoes. But should we be? Well, Rachel certainly hasn't done the right thing here. It is Amy's first day and it isn't nice to bad-mouth someone in that way. The question is, though, despite this transgression, should Amy forgive Rachel and give her another chance?

Yes. Yes. Yes. She should! She has to! We have to forgive these digressions. And we shouldn't just forgive them once, we should forgive them multiple times. We should forgive people for making silly comments, for calling us names behind our backs and for not noticing that we are there.

And why should we forgive them? Because we are being nice? Because we let ourselves get trodden all over? Because we are weak and don't stand up for ourselves?

No. No. No. That's not it at all. We should forgive them because we are rational, because we believe in logic and reason. We should forgive them because we want to be fair. We should forgive them because of what we have learnt from Reverend Bayes. We should forgive them because the second equation tells us that it is the *only* right thing to do.

Here is why. Bayes' rule is the connection we need to make between model and data. It allows us to check how well our cinematic pictures correspond to reality. In the example at the start of this chapter, I worked out the probability $P(CRASH \mid SHAKE)$ that the plane would crash given it was shaking violently. Amy wants to know $P(BITCH \mid DISS)$ and the logic is just the same.

The CRASH and BITCH are the models inside our heads. They are our beliefs about the world and take the form of thoughts or, in my case, cinematic films. The SHAKE and the DISS are the data we have access to. Data is something tangible, something that happens, something we experience and feel. Much of applied mathematics involves reconciling models with data, confronting our dreams with the harsh truth of reality.

We write M for model and D for data. What we want to know now is the probability our model is true (Rachel is a bitch) given the data (rude comment in toilet) we have:

$$P(M|D) = \frac{P(D|M) \cdot P(M)}{P(D|M) \cdot P(M) + P(D|M^C) \cdot P(M^C)}$$

(Equation 2)

To understand this equation – Bayes' rule – it is best to break the right-hand side of the equation into parts.

The numerator (the top line of the equation) involves multiplying two probabilities, $P(M)$ and $P(D|M)$. The first of these, $P(M)$, is the probability that a model is true before anything happens, the statistical probability of plane crashes or Amy's estimate of the probability that the people she meets are bitches, that is, 1 in 20 for the latter probability. This is what Amy knows before she goes into the bathroom. The second probability, $P(D|M)$, is about what happens in the bathroom. It is the probability of Rachel dissing Amy in the bathroom if she really is a bitch or, more generally, the probability we observe data if the model is true. This is a hard one to put a number to, but let's assume it to be a toss-up: $P(D|M) = 0.5$. Even if Rachel is a bitch, she doesn't spend every bathroom trip bad-mouthing her fellow students. Bitches spend at least 50% of their time talking about something else.

The reason we multiply the two probabilities, that is, $P(D|M) \cdot P(M)$ in the numerator is to find the probability that both are true. For example, if I throw two dice and want to know the probability that they are both six, then I multiply 1/6 for the first die being six, by 1/6 for the second die, and I get $1/6 \cdot 1/6 = 1/36$ as the probability of two sixes. The same multiplication principle applies here: the numerator is the probability Rachel is a bitch *and* she made bitchy comments on a visit to the bathroom.

While the numerator of Equation 2 deals with Rachel being a bitch, we also have to consider an alternative model in which Rachel is nice.

We do this in the denominator (the bottom of the equation). Rachel could be a bitch making a bitchy comment (M) or a nice person making a mistake (M^C). The superscript C in M^C stands for complement. The complement in this case is being 'nice'. Notice that the first term in the denominator, $P(D|M) \cdot P(M)$, is just the same as the numerator. The second term, $P(D|M^C) \cdot P(M^C)$ is the probability that Rachel makes a nasty comment given that she is *not* a bitch, multiplied by the probability that people typically are nice. By dividing through by the sum of all eventualities we cover all potential explanations of the data Amy observed in the cubicle, thus giving $P(M|D)$, the probability of the model given the data.

If Rachel isn't a bitch, then she is nice, so $P(M^C) = 1 - P(M) = 0.95$. We now need to consider the probability that nice people make mistakes. Rachel might very well be a good person who is having a bad day – we have all had them. Let's set $P(D|M^C) = 0.1$, to denote the probability that 1 day in 10 a nice person can have a bad day and say something they might regret later on.

All that's left to do now is the calculation, which is illustrated in Figure 2. This is done in exactly the same way as in the aeroplane crash example, but with different numbers:

$$P(M|D) = \frac{0.5 \cdot 0.05}{0.5 \cdot 0.05 + 0.1 \cdot 0.95} \approx 0.21$$

The probability that Rachel is a bitch is around 1 in 5. This is the reason Amy should forgive Rachel. There is a 4 in 5 chance she is a nice person. It would be totally unfair to judge her on this one act alone. Amy shouldn't mention what she heard Rachel say, or let it affect how she interacts with her. She should just wait and see how it works out tomorrow. There is an 80% probability that the two of them will be laughing together about the toilet cubicle incident by the end of the college year.

There is one further word of advice to Amy, cowering behind the toilet door. Maybe she wasn't at her best that morning when she heard Rachel bitching about her in the bathroom. She probably could have concentrated a bit harder when they were working together and,

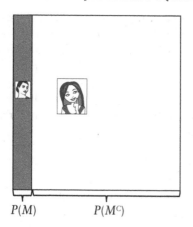

Before we have data, we already have a model, *M*, of the world.

The areas of the rectangles are the probabilities $P(M)$ and $P(M^c)$ that our model is true or false.

$P(M)$ $P(M^c)$

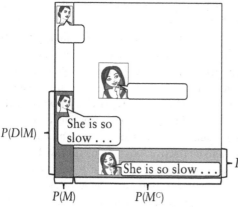

Now decide the probability of data given to each model: $P(D|M)$ and $P(D|M^c)$.

One of the two shaded rectangles contains the true state of the world.

$P(D|M)$

$P(D|M^c)$

$P(M)$ $P(M^c)$

$P(M|D)$ is found from the relative size of the shaded rectangles, using Equation 2.

Figure 2: Illustration of Bayes' theorem

let's face it, she shouldn't be sitting on the loo and fiddling with her phone after lunch. But remember, Bayes forgives transgressions. She should apply the same rule to herself that she applies to Rachel. Bayes' rule tells her to slowly adjust her opinion of herself and not be too discouraged by specific events.

You are the product of all your actions, not the result of one or two mistakes. Apply to yourself the same rational forgiveness that Bayes asks you to apply to others.

*

The first lesson to draw from Bayes' rule, the judgement equation, is that we should be slow to draw definitive conclusions. The numbers I used in the example do affect the result, but they don't affect the logic. You can try yourself: what proportion of people do you think are generally nice? How often do nice people make mistakes? And how often do bitches do bitchy things? Put your own numbers into the equation and you will reach the same conclusion: it takes more than one nasty comment before you should label someone a 'bitch'.

Sometimes my boss acts like an asshole. Sometimes my students seem to lack focus. Sometimes one of the researchers I work with wants to take credit for something I consider as my idea, claiming that he thought of it first. Sometimes the chair of the committee I am on appears disorganized and ineffective, wasting my time with pointless email exchanges. In these situations, I use the judgement equation. That doesn't mean that I create calculations to assign a probability that each of my colleagues is an asshole, or unfocused or incompetent. What it means is that I try not to let individual events determine how I feel. If I perceive that someone I work with has made a mistake, then I wait to see how the situation develops. It could very well be the case that it is me who is in the wrong.

In *Pride and Prejudice*, Mr Darcy tells Elizabeth Bennet that his good opinion once lost is lost for ever. In reply, Miss Bennet remarks that, 'Implacable resentment is a shade in a character.' How carefully and correctly worded Jane Austen's phrasing is! Even in criticizing Darcy, Miss Bennet restrains herself enough to see his resentfulness as only a passing shade and not a deep stain. It is this care, taken in the formation of our opinions of others, that is the sign of excellent judgement.

*

We cannot understand TEN without unpacking its history and its philosophy. The story of TEN is about a small group of people who have passed the secrets of rational thinking itself from generation to generation. They have asked big questions. They wanted to know how to think more clearly and more precisely. They wanted to be able to evaluate the truth of what we and others say. They have even asked what it means for something to be true or false. Theirs is a story about the really big issues: the nature of reality and their place in it.

It is also a story about religion, and about right and wrong. It is a tale about morality, of both good and evil.

The first stopping point in our story is 1761. Dr Richard Price has just found an essay among the papers of a recently deceased friend. The document contains a combination of mathematical symbols and philosophical reflections. It asks the reader to consider 'a person just brought forth into this world [who] collect[s] from his observation of the order and course of events what powers and causes take place in it'. It asks how such a person should reason after seeing his first sunrise, his second sunrise and his third. What should he conclude about the probability that the sun rises each day?

The essay drew a remarkable conclusion. The daily sunrise should *not* lead our newly arrived man to believe that the sun will always rise. Rather, it entreated him to be very cautious about the arrival of the sun, even after seeing one hundred sunrises, and even after a lifetime of sunrises. Nothing should be taken for granted.

That friend, and the author of the essay, was Thomas Bayes. He set out how to estimate, from the data of previous occurrences, the probability that an event will occur again. Bayes instructed the 'brought forth' man to represent his estimate of the probability of a sunrise occurring using a parameter θ. Before the first sunrise, the man should not have any prior conceptions about the sun and should thus consider all values of θ to be equally likely. At that point it is just as conceivable that the sun rises every day ($\theta = 1$), as that it rises on half of the days ($\theta = 0.5$) or that it only rises one day in a hundred ($\theta = 0.01$). Although θ is somewhere in between o and 1 (all probabilities have to be less than or equal to one) there remains an infinite number of values it could take. For example, it could be 0.8567, 0.1234792, 0.99999 and so on. The decimal points can continue to any level of precision, provided the value for θ is between 0 and 1.

To handle the problem of precision, Bayes instructed the man to decide what he considered the smallest plausible probability of a daily sunrise. If he thought there was at least a 50% chance of a sunrise each day, then he should write that $\theta > 0.5$. If he thought there was a greater than 90% chance of a sunrise he should write $\theta > 0.9$. Imagine now that after seeing 100 sunrises, the man proclaims that he believes that the sun rises on more than 99 out of 100 days. In doing so, he estimates that $\theta > 0.99$. Writing P($\theta > 0.99$ | 100 sunrises) expresses the probability that he is correct in his estimation. Bayes showed, using a version of Equation 2 that accounts for differing precision levels, that P($\theta > 0.99$ | 100 sunrises) $= 1 - 0.99^{100+1} =$ 63.8%.[2] There is a 36.2% chance that the man has got it wrong and the sun rises less often than he believes.[3]

If the man lived for sixty years and saw a sunrise every day, then he could be certain that the chance of a daily sunrise exceeds 99%. But if he claimed that sunrises happen on more than 99.99% days, we would still have to advise some caution. $1 - 0.9999^{365\times60+1} = 88.8\%$. There remains an 11.2% chance that our man has got it wrong. Bayes forces the newly arrived inhabitant of the world to state his model, to say what he thinks is the minimum possible value of θ and then tells him the probability that he is correct in his assumption.

Richard Price realized that Bayes' equation was related to a debate during the eighteenth century about miracles. Price was, like Bayes, a minister in the Church and was interested in how the new scientific discoveries of the time could be squared with the miracles he believed in from his study of the Bible.

A decade earlier, the philosopher David Hume had argued that 'no testimony is sufficient to establish a miracle unless the testimony be of such a kind that its falsehood would be more miraculous than the fact which it endeavours to establish'.[4] Hume's argument can be viewed as an appeal to the judgement equation. It asks us to compare the model M that miracles occur against the model alternative M^C that miracles don't occur. Hume argues that, since we have never witnessed a miracle before, $P(M^C)$ is nearly one and $P(M)$ is very small. We would thus need a very substantial and convincing miracle, one with a very high $P(D|M)$ and low $P(D|M^C)$ to convince us otherwise. Hume's argument is similar to the argument I made at the start of the chapter about a shaking plane: we would need a very strong piece of evidence in order to convince us that an otherwise reliable plane is going to crash. We would need a very strong piece of evidence to convince us that Jesus came back from the dead.

Price found Hume's reasoning to be 'contrary to every reason'.[5] Hume had misunderstood Bayes. He explained that Hume had to be more precise about what he was saying about θ, the probability of that a miracle doesn't occur.[6] Even those who believe in miracles don't believe that they happen every day. To make the argument more concrete, imagine Price forces Hume to state his belief in the frequency of miracles and he says that they occur less often than once in about every 10 million days or 27,400 years, giving $\theta > 99.99999\%$.[7] Let's say that Price believes that $99.99999\% > \theta > 99.999\%$, that miracles occur less often than once every 274 years but more frequently than once in 27,400 years. Now imagine that no miracle has occurred for 2,000 years. The probability that Hume is correct, given the data, is about 7.04%. The probability Price has got it right is 92.89%. Even if no miracle occurs over several millennia, the evidence isn't strong enough to dismiss miracles. There simply isn't enough data in one person's lifetime to support Hume's claim that miracles don't happen.

Richard Price set TEN on a path of Christian morality. He believed in the resurrection of Christ and he used rational argument to cast doubt on doubt. Price believed that logical thinking could reveal truths about the world hidden from our everyday experience. God was one of these truths.

Two millennia earlier, in his allegory of the cave, the Greek philosopher Plato had described uncritical humans as chained within a cave looking at shadows, only able to see confusing projections of a truer, more logical world on the outside. Plato's allegory is often used as a way of explaining the power of mathematics, and it was an allegory that Price took very seriously. He believed that we discover new truths by accepting that the shadows on the cave wall are not reality. Our everyday experience is a messy representation of a greater truth. By thinking more clearly about the true form of the world – by means of models that are independent of data – we can think more clearly about messier situations, the shadows of our everyday lives.

The TEN that Price envisaged was formed from his religious beliefs and Plato's metaphysics.[8] He believed there was a morality in mathematics, that there was a rational, correct way to approach life. Not only did he preach this message, he put it into practice. By compiling tables of life expectancies, he provided new payment schemes that were used for nearly a century afterwards in life assurance.[9] He saw his work as a way of protecting the poor from uncertainties, showing that nearly all the assurance companies at the time could not meet their future obligations and needed to improve their policies.[10] Price was an ardent supporter of the American Revolution and a close friend of Benjamin Franklin, seeing opportunities in America to create a system based on principles of liberty, with equality in land ownership and in which political power was fairly distributed among all people.[11] The USA was, in Richard Price's vision, to be the country where a religious, rational TEN could finally flourish.

TEN's modern practitioners seldom talk about morality, and only a minority believe in a Christian God, but many have inherited Price's values: the actuary carefully calculating your father-in-law's car insurance premiums; the government official planning our pensions or setting interest rates; the staff scientist at the United Nations setting development goals; the climate scientist measuring the probability of different temperature increases over the next twenty years; or the healthcare professional balancing the risks and costs of medical treatments. They use Bayesian judgement to create a more orderly, fairer and better structured society. They help us share risks and uncertainty, so that when a terrible but rare event befalls one of us, the contributions made by the rest of us cover the costs.

The judgement equation leads its members to act for the benefit of all. Good judgement, seen through the eyes of Price, demands that we are both forgiving and considerate of others. It tells us that we shouldn't dismiss miracles. It suggests that at least one of the Ten Equations puts us on a path to righteousness.

The audience sits quietly, waiting for the day's events to commence. I can see on Björn's face that he is nervous. He has spent the last five years of his life dedicated to the lofty goal of discovering new truths, to doing academic research. I have been his PhD supervisor, guiding him towards his goal. Now, standing in front of his peers, colleagues and an examination committee, but also his friends and his family, it is time for him to defend his doctoral thesis.

It is the combination of this varied audience and the challenging subject area of his research that makes Björn nervous. One of the chapters in his thesis is entitled 'Last Night in Sweden', and is a study of the connection between violent crime and immigration in his homeland. In another chapter, he looks at how the Swedish Democrats, a populist anti-immigration party, has risen to prominence over the last ten years in a country famed for its liberal and socialist policies.

To the mathematicians on the examining committee and sitting in the audience, this is a thesis about statistical methods. To his co-supervisor, Ranjula Bali Swain, an economics professor who works on everything from sustainable development to how microfinance lifts women out of poverty, Björn's thesis aims to explain what happens as cultures mix around the world. Björn's family, the Blomqvists, and his friends want to know what he has found out about a Sweden in flux. Their country is changing from a homogeneous land of Vikings to a multicultural melting pot of Afghans, Eritreans, Syrians, Yugoslavs and Brexited British.

Björn is anxious that he will fall off the tightrope he is balancing on in order to make everyone happy. The defence of PhDs in Sweden starts with the opponent, the person charged with reading and discussing the thesis with the candidate, presenting the background to the research area. Björn's opponent is Ian Vernon from Durham University.

Ian talks us through the principles of Bayesian reasoning. While my examples in this chapter focus on testing just one model or one parameter, scientists usually have many different competing hypotheses. The challenge for Ian is to search through all these alternative models and assign each of them a probability. No hypothesis is 100% true, but as evidence accumulates, some become more plausible than others. He builds up through examples, starting with a search for oil reservoirs. A patented algorithm developed by Ian and his colleagues is used by oil companies to find the reserves that offer the best long-term prospects. Then he moves to health. When researchers trial an intervention to eradicate malaria or HIV, they first create a mathematical simulation to predict the effect of their intervention. The Bill and Melinda Gates Foundation is using Ian's method to help plan their disease-eradication programmes.

Finally, Ian moves on to one of the biggest questions of all. What happened in the early stages of our universe? How did galaxies first form after the big bang and what models explain the size and shape of those which we can observe today? Ian was able to narrow down the number of possible models of the early universe, finding likely values for seventeen different parameters which decided how galaxies expanded into space.[12] Ian's presentation balances the audience tightrope perfectly, showing the power of mathematical methods and a broad range of applications. Björn's family and friends gasp as they watch a simulation of galaxies spinning and colliding, a possible model of the early time evolution of our universe, the parameters for which had been recovered using Reverend Bayes' rule.

Now it is Björn's turn to present his work. An introduction on the scale of the universe could easily have overwhelmed an already nervous PhD student. Björn might have worried how his own study of one country in Scandinavia compared to the sheer scope of Ian's research. When I look at Björn, though, I can see that he is now relaxed and ready. And when I look back to the audience, at his parents, I can see the pride on their faces. *This* was what could be done with the mathematics that Björn had been learning, the Blomqvists thought. *These* were the skills that their Björn was mastering: the mathematics of the universe.

Societal change is just as complex as the origins of the universe, albeit in a very different way. Björn shows how the rise of the anti-immigration Swedish Democrats can be explained primarily by geographical location. Certain regions, particularly in the southern-most region of Skåne, but also in the central area of Dalarna, support the Swedish Democrats. Surprisingly, these aren't areas where immigration is highest. It isn't because immigrants are moving into an area that resentment occurs. Rather it is in rural areas, particularly where people have lower levels of education, that support for an anti-immigration agenda has been growing.

After Björn has finished his presentation, he is interrogated by Ian and the thesis committee. Ian and other mathematicians on the committee want to know the technical details of how Björn had compared models with data. Lin Lerpold, an economist colleague of Ranjula and a member of the committee, points out some important limitations of his study. Björn hasn't fully got to the bottom of the causes of anti-immigration sentiment. He has looked at patterns of change within local communities, but he hasn't understood the minds of the individuals living in these communities. In-depth interviews and questionnaires would be needed to answer Lin's questions.

The committee's interrogation was tough but fair, and their verdict was unanimous. Björn had passed his viva. He had joined the elite force of Bayesian scientists.

*

Bayesian reasoning has transformed how we do science and social science over the last few decades. It fits perfectly with the scientific way of seeing the world. Experimentalists collect data (D) and theoreticians develop hypotheses or models (M) about that data. Bayes' rule puts these two components together.

Consider the following scientific hypothesis: *Mobile phone usage is bad for teen mental health.* This is a hotly debated issue in my household, with two teenagers (and, to be fair, two adults) glued all day long to their screens. When I was young, my parents worried where I was and what I was up to. My wife and I don't have that problem. We worry instead that our kids spend too much time sitting inside staring at the soft blue light of their mobile phones. What we wouldn't give for a good old-fashioned 'Why didn't you come home on time and who were you out with?' argument . . .

Dr Christine Carter, sociologist and author of several self-help books about parenting and productivity, has come down very firmly against too much mobile phone usage, writing that 'screen time is a likely cause of the ongoing surge in teen depression, anxiety, and suicide'. Her article, for *Greater Good Magazine* at the University of California, Berkeley, makes the argument in two steps.[13] First, Carter references a survey of parents, nearly half of whom believed that their teenage children were 'addicted' to their mobile device and 50% of whom were concerned that it was negatively affecting their mental health. Her second step was to reference data from a study of 120,115 adolescents in the UK who answered fourteen questions about how they felt, in terms of happiness, life satisfaction and about their social lives. The study found that, above a lower threshold of one hour a day, children who spent more time on their smart phones had lower mental well-being, as measured by the questionnaire. In other words, the more children use their mobiles the unhappier they are.

Sounds convincing, doesn't it? I have to admit that when I first read the article, I was persuaded. Written by a researcher with a PhD, published in a magazine based at a world-leading university, the article uses peer-reviewed science and rigorous survey data to support its case. But there is a problem and it is a big one.

Christine Carter has only filled in the top part of the judgement equation. Her first step, to describe parents' fears, is analogous to $P(M)$, the probability a parent believes that screen time affects mental well-being. Her second step is to show that the current data is consistent with the worried parents' hypothesis, i.e. $P(D|M)$, and is reasonably large. But what she has neglected to do is to account for other models that could explain the well-being of modern teenagers. She has calculated the numerator (top part) of Equation 2, but neglected to tell us about the denominator (the bottom part). Carter hasn't told us $P(D|M^C)$ for the alternative hypotheses, and thus leaves us none the wiser as to $P(M|D)$, the probability that mobile phone usage explains teenage depression, which is what we actually want to know.

Candice Odgers, Professor of Psychological Science at the University of California Irvine, has filled in the blanks left by Carter. And, in a commentary published in the journal *Nature*, she reached a very different conclusion.[14] She begins her article by acknowledging the problems. In the US, the proportion of girls aged between twelve and seventeen reporting depressive incidents has increased from 13.3% to 17.3% from 2005 to 2014, and a smaller increase has also been seen for boys of the same age. There is little doubt that mobile phone usage has increased over the same period; we don't need statistics to agree on that. Odgers also didn't dispute the data from the study of UK adolescents – referenced by Christine Carter – that indicates an increase in cases of depression among heavy phone users.

What Odgers pointed out, though, was that other hypotheses can also explain depression among young people. Not eating breakfast on a regular basis or not sleeping the same hours each night were both three times as important as screen usage in predicting poorer mental well-being.[15] In the language of Bayes' theorem, breakfast and sleep are alternative models that could explain depression, and these have a high probability, $P(D|M^C)$. When these models are put into the denominator of Bayes' rule, they outweigh the numerator, and $P(M|D)$, the probability that mobile phone usage is related to depression, becomes smaller – not completely negligible, but small enough that it fails to offer an important explanation of teenage mental-health issues.

There is more. You see, there are also documented benefits for teenagers who use mobile phones. A large number of studies have shown that kids use phones to offer each other support and build long-lasting social networks. For most middle-class kids, the group which is usually targeted by those offering advice about screen time, mobile phones improve their ability to make real and lasting friendships, not just online but in real life too. The problems, Candice Odgers went on to show in her article, are for kids from disadvantaged backgrounds. Less well-off adolescents are more likely to get into fights over something that has happened on social media. Kids with a history of being bullied in real life are later more likely to be victimized online.

My kids connect with people all over the world and they often learn about new ideas online. I overheard Elise and Henry discussing bongo drums and cultural appropriation the other week.

Elise said, 'It's just basic respect that if someone tells you they are offended by you playing their culture's music, then you should stop.'

'Is Eminem cultural appropriation then?' countered Henry.

There is no way I would have had a discussion like that with my sister when we were thirteen and fifteen. I'm not even sure that we could carry it off today. Kids born in the 2000s have access to important ideas and information that would have been beyond the comprehension of those of us growing up in the 1970s, 1980s or even the 1990s.

*

I want to go back to Amy and Rachel, because there is something I skipped over, and it is important.

The numbers I used in the example – 1 in 20 people are bitches; bitches spend 50% of their time bitching and even nice people have one off day in every 10 – are not only slightly arbitrary, but they are also subjective. What I mean by subjective is that they vary between people. Depending on your particular life-experiences, you will trust people more or less than Amy does. We can contrast this with aeroplane crashes: they are a terrible and an objective reality. Amy's choice in how she sees new classmates or how I categorize my co-workers is entirely based on our own subjective experience of the people we have met before. There is no objective measure of bitch-iness or asshole-ness.

It is true that the numbers in the Amy story are subjective. But here is the thing. Bayes' rule doesn't just operate on objective probabilities, it operates on subjective ones too. As long as we can give numbers, and these don't have to be completely accurate, then Bayes' rule allows us to reason about these numbers. While we can change the numbers and get different results, what can't be changed is the logic Bayes tells us to apply.

These assumptions are called *priors*. In Equation 2, $P(M)$ is the prior probability that our model is true. In many cases, prior probabilities can be derived from subjective experience. What can't be subjective is how we then determine $P(M|D)$, the probability our model is true after we have seen data. This calculation must follow Bayes' rule.

Many people think of maths as being all about objectivity. It's not. It is a way of representing and arguing about the world, and sometimes the things we argue about are known only to us. In the end, no one else might ever know or care whether or not Amy thinks Rachel is a bitch. The whole thought process could be hidden in her brain for ever.

Think back to my cinematic representation of the world – all these films I play to myself, day in day out, some of which are very personal. They might be concerns about how my wife feels, or thoughts about my daughter's future; or they might be a movie where I lead my son's Futsal team to victory, culminating in the winning of a cup, or dreams about how I might one day be a bestselling author. I don't need to tell you about them, because they belong entirely to me. The judgement equation doesn't tell us what films we should have in our collection or what we should dream about. It just tells us how we should reason about these dreams, because each of these 'films' is a model of the world. The judgement equation allows us to update the probabilities we associate with each dream, but it doesn't tell us which dreams to have.

'Many people, even mathematicians and scientists, fail to realize that the real power of Bayes lies in how it forces you to reveal the way you thought before you did an experimental study and afterwards,' Ian Vernon told me over a glass of champagne after Björn's PhD defence. 'A Bayesian analysis demands that you break down your argument into different models and look for support for these one at a time. You might think that the data supported your claim, but you need to be honest about how much support you gave your hypothesis before you did the experiment.'

I agreed. Ian was talking in general, thinking back to Björn's defence and his use of Bayes to explain the rise of the extreme right in Swedish politics. This was a project in which I had worked through all the details with Björn, learning about all the factors that lead people to vote for nationalistic parties. Now I was trying to apply the same approach to a question in my family life. I am not an expert in mental health or mobile phones, but the judgement equation offers me a way of interpreting the research results obtained by others, a way of assessing the relative merits of the arguments put forward by scientists. I used Bayes' theorem to check that each of them had fulfilled the criteria for good judgement. Had the researchers looked at both their own and alternative models of the world? Candice Odgers balanced all sides of her argument; Christine Carter presented only her own side.

I am sometimes disappointed when I see how uncritically advice from so-called experts on parenting, lifestyle and health is received. Much like the uninformed gamblers who message me for tips about the upcoming big game, the consumers of this advice see only as far as the latest study. They don't realize that developing a healthy and well-balanced lifestyle involves taking a long-term position, just as successful gambling involves a long-term strategy.

However, it is not entirely Christine Carter's responsibility to make sure that she presents all sides of the story. You might think it strange that I take this position now, given that I found her work misleading, but I also realize that she is reflecting worries held by many parents, including myself. The data she describes is real and she proposes support for her model. It isn't completely up to her to describe support for alternatives.

It is, in big part, *our* responsibility to check the validity of her model. When I read opinion pieces, I check that the authors, no matter their credentials, have all the pieces of the equation in place. It wasn't hard for me, as a parent, to gain a fuller picture of the role of screens in our lives. All the articles I used are freely available online and it took a couple of evenings of my own screen time to download and read through them. Once I understood the debate, I then discussed the results with my teenage children. I told them that a good night's sleep and eating breakfast were both three times more important for their mental well-being than the amount of time they spent on their phones. I discussed what this meant, emphasizing that it didn't mean they should spend every evening lying on the couch watching YouTube. Exercise and social interactions were also important and they should definitely not have their phones in the bedroom. I think Elise and Henry understood.

The same people who uncritically consume the information provided to them by parenting tipsters can be heard expressing scepticism when they hear other scientists, like Candice Odgers, who take a more balanced view. When scientists present all sides of an argument, they are sometimes viewed as being unsure about their conclusions.

Topics like climate change, the merits of different diets and the causes of crime are actively debated within the academic community. Such discussions, and a comparison of all potential hypotheses, are not a sign of weakness or indecisiveness on the part of those involved in these debates. Rather, it is a sign of strength and thoroughness. It is a sign of having an edge, of having considered all possibilities.

*

The world is full of people offering advice. How to be organized at work and in the home. How to stay calm and focused. How to be a better person. Choose your perfect job. Choose your perfect partner. Choose your perfect life. The top ten things to do when starting at a new workplace. The top ten things to avoid. The top ten equations.

Yogic calmness. Open mindfulness. Thinking deeply and breathing slowly. Tigers. Cats and dogs. Popular psychology and evolutionary behaviour. Be a caveman, a hunter-gatherer or a Greek philosopher. Switch off. Connect. Wind down. Power up. Stand up straight and never lie. Eat the right things and you'll never die. Don't give a f*ck and you'll always be happy. Do it now and make it snappy.

All this advice lacks structure. The important information is mixed up with opinions and nonsense. The judgement equation allows you to organize and evaluate. It turns each piece of advice, wanted or unwanted, into a model which can be tested against data. Listen carefully to other people's opinions, list the alternatives, collect the data and make a judgement. Adjust your own opinions as evidence slowly accumulates for and against ideas. Use the same process when you judge the actions of others. Always give them a second chance, and even a third, making sure you let the data, not your emotions, lead your decisions. By following Bayes' rule, not only will you make better choices in life, but you will also find that you gain the trust of others. You will become known for your good judgement.

3

The Confidence Equation

$$h \cdot n \pm 1.96 \cdot \sigma \cdot \sqrt{n}$$

Not all of TEN was born out of Christian morality. If we were asked to travel to a single place and point in time when TEN was established, then that wouldn't be Thomas Bayes' deathbed: we would still find ourselves in London, but nearly thirty years earlier, at a gathering of friends on 12 November 1733, and we would be listening as Abraham de Moivre revealed the secrets of gambling.

De Moivre was a streetwise mathematician. He had been exiled from France for his Protestant religion and was treated with suspicion in London for his French origins. So, while contemporaries like Isaac Newton and Daniel Bernoulli became professors in their fields, de Moivre was forced to find other ways to support himself. Part of his income came from tutoring middle-class boys in London (it has been speculated, although not proven, that a young Thomas Bayes was one of his pupils) and the rest was obtained from 'consultancy'. He could be found in Old Slaughter's Coffee House on St Martin's Lane giving advice to everyone from gamblers and financiers to Sir Isaac Newton himself.

The work de Moivre presented in November 1733 was a step up in sophistication from his earlier writing. It showed how the new mathematics of calculus, recently developed by Newton, could be used to determine our confidence in the long-term profitability of games of chance. Ultimately, the equation he presented was to become the basis for the way scientists and social scientists were to establish confidence in their research results. But in order to understand where this confidence equation comes from, we first need to begin in the same place as de Moivre. We need to enter the shady world of gambling.

*

Nowadays it takes about two minutes to open a betting account at an online casino. Name, address and, most importantly, credit card details and you are all set. The games vary. There is online poker, where you play against others and the house takes a cut. There are slot machines, much like those that used to be found in pubs; they have names like Cleopatra's Tomb, Fruit vs Candy and Age of the Gods, as well as trademarks like Batman v Superman and Top Trumps Football Stars. You press a button, spin a wheel and if the Gods align or enough Batmen appear in a row you win. Finally, there are traditional casino games like blackjack and roulette streamed via live videos where smartly dressed young men deal the cards and women in low-cut evening dresses spin the roulette wheel.

I opened an account with a leading online gambling site. I put in £10, collected a £10 new account bonus, giving me a starting capital of £20. I decided to start with some Age of the Gods, for no other reason than it allowed me to make smaller bets than any other game. At 10p per spin, I was guaranteed more spins.

Twenty spins later I was down by 70p and it didn't feel like anything was happening. I was bored of the Gods, so I went into Top Trumps and started spinning Ronaldo, Messi and Neymar. It was more expensive, at 20p per spin, but after 6 spins I got a big win: £1.50! Now I was nearly back to my original starting capital. I had a go at Batman v. Superman and a few others. Then I discovered the auto-spin setting so that I could spin repeatedly without having to keep pressing a button. Not a good idea. Two hundred spins later, I was down to £13.

The slot machines didn't feel like value for money, so I decided to try the live casino instead. The table was run by Kerry, a woman in her twenties in a black dress. She welcomed me into the room and was already chatting with another customer. It was a strange experience. I could type her messages and she would answer. 'How is the weather where you are?' I asked.

'Good,' she replied, looking straight at me, 'feels like spring is on the way. Last bets now. Good luck to you.'

She was based in Latvia and was surprisingly open, telling me she had been to Sweden four times before. After a bit of chitchat, I asked her if there had been any big wins today.

'We don't see how much you are betting,' she told me.

I felt a bit silly. I'd been making sure to place a £1 bet on every spin of the wheel so she didn't think I was cheap.

I liked Kerry, but I felt I had to look around a bit more. I don't know quite how to put this, but there was a reason she, and most of her male colleagues, were in the lower stakes room. She was slightly awkward and uncomfortable in her tightly fitting dress. Kerry wasn't sexy.

The higher stakes rooms were different. The cut of the dresses was lower and the smiles more inviting. Before each spin, Lucy, whose high-stakes room I was now in, would look up at the camera knowingly, as if to say that the specific choice I had made was the right one. I had to force myself to remember that she wasn't just looking at me, but also at 163 different gamblers from all over the world.

She was answering questions fielded by her clients. 'Yes, I have a partner. But it's complicated,' she was telling one.

'Oh, I love travelling,' she told another. 'I'd love to go to Paris, Madrid, London . . .'

The camera cut again to a top-down view, providing a glimpse of her legs before the wheel was spun.

I started to feel very uncomfortable. I needed to remind myself why I was here in the first place. I went back into a lower-stakes room with Max, a polite young man who gave statistical advice on the colours and numbers that had been winning. High numbers were doing well on his wheel, apparently.

I studied my account balance. I had been playing red and black at random, without thinking too much, and was surprised to find that after a couple of hours in the casino I now had £28. Up £8 for the evening. Things were going well.

How do we know if we are winning because we are skilful or just because we are lucky? At the online casino I knew the odds were stacked against me, even though, after a couple of hours' play, my balance was larger than when I had started.

For other games, I don't know whether I have got an edge or not. If I play poker with friends, I see my pile of chips go up and down, but how long until I can safely say I am the better player? If I set up a sports-betting strategy the way I did for the World Cup, when do I know that it is paying off?

These questions aren't limited to games and gambling; they can be political too. How many voters do we need to poll before we can accurately estimate who will win the US presidential election? They can be about our society: how do we know if a company is guilty of racial discrimination when hiring people? The questions can even be personal: how long should you give your job or a relationship before you decide to make a change?

Amazingly, there is a single equation that answers all of these questions: the confidence equation. Here it is:

$$h \cdot n \pm 1.96 \cdot \sigma \cdot \sqrt{n}$$

(Equation 3)

The concept of confidence is captured by the central symbol of the equation, the ±, plus or minus. Imagine you ask me how many cups of coffee I drink per day. I don't know for sure, so I might say 'four plus or minus a couple', 4 ± 2. This is a confidence interval, convenient shorthand for giving both the average and an idea of the variation around that average. This doesn't mean I never drink seven cups (or only one), but it means that I am reasonably confident that most days I drink between two and six cups.

Equation 3 allows us to make more precise statements about our confidence. Imagine I asked all of the readers of this book to play the roulette wheel $n = 400$ times, staking £1 with each spin on either black or red. The roulette wheel has 37 numbers: 1 to 36, which are arranged in an alternating pattern of red and black; and the green number zero which is special. It biases the wheel in the casino owner's favour. If, for example, a gambler bets on red, then the probability that the ball lands on red and he doubles his money is 18/37. The probability he loses his money is 19/37. The gambler's expected (average) profit/loss per £1 spin is $1 \cdot 18/37 - 1 \cdot 19/37 = -1/37$; that is, an average loss of approximately 2.7p per spin. In Equation 3, the average loss is denoted as h and in this case $h = -0.027$. Over 400 spins, each of my readers can expect to lose $h \cdot n = -0.027 \cdot 400 = 10.8$ dollars on average.

The next step is to calculate the level of variation around our average loss. Not every reader will lose (or win) the same amount. Even without doing arithmetic we can see that there is a large variation in the possible outcomes of every spin of a roulette wheel: if I bet £1, then I have either doubled my money or lost it after the bet. The variation for one spin is about the same size as the investment and is much larger than the average loss of 2.7p.

We can quantify this variation more by taking the square of the distances of the outcome of any one spin to the average loss per spin. The square of the distance from a win of £1 to the £0.027 average loss is $(1 - (-0.027))^2 = 1.0547$ and the square of the distance from a loss of £1 to the average loss is $(-1 - (-0.027))^2 = 0.9467$. Since the win outcome occurs 18 occasions out of 37 and the lose outcome occurs on 19 occasions, then the average squared distance, denoted σ^2, for a single spin of the roulette wheel is

$$\sigma^2 = \frac{18}{37} \cdot 1.0547 + \frac{19}{37} \cdot 0.9467 = 0.9993$$

This average squared distance, σ^2, is referred to as the variance. The variance in roulette is very close to, but not exactly, 1. If the roulette wheel were fair, if it had just had 36 numbers, half of which were red and half black, then the variance would be exactly 1.

The variance increases in proportion to the number of spins we make. If I spin the roulette wheel twice, then the variance doubles; if I spin it three times the variance triples and so on. So, the variance of n spins is $n \cdot \sigma^2$.

Notice that, because we squared the distances between the outcomes and the averages, the unit of the variation is dollars squared, rather than dollars. In order to get back to dollar units, we take the square root of the variance, to give the standard deviation, denoted σ, which in this example is 0.9996. The square root of n is written $\sqrt{n}$. Thus, the average plus or minus the standard deviation for the money won or lost from 400 roulette win spins is

$$\sigma \cdot \sqrt{n} = 0.9996 \cdot \sqrt{400} = 0.9996 \cdot 20 = 19.99$$

We now have most of the components of the confidence equation. The only part of Equation 3 we haven't yet explained is the numerical value 1.96. This number comes from a mathematical formula which today we call the Normal curve, the bell-shaped curve that is commonly used to describe the distribution of our heights and our IQs. You can visualize the Normal distribution as a bell with a peak at the average value (at −10.8 for 400 spins of the roulette wheel or 175cm for men's heights in the UK).[1] This Normal bell curve is illustrated in Figure 3 for £1 bets on red or black for 400 spins of the roulette wheel.

Now imagine we want our interval to enclose 95% of the bell-shape. For 400 spins of the roulette wheel this is the interval of wins or losses in which 95% of the reader's profits or losses occur. The value of 1.96 comes from this interval. In order to contain 95% of the observations we need to make our interval 1.96 times larger than the standard deviation. In other words, the 95% confidence interval for our profit after 400 spins of a roulette wheel is Equation 3, that is

$$h \cdot n \pm 1.96 \cdot \sigma \cdot \sqrt{n} = -0.027 \cdot 400 \pm 1.96 \cdot 0.9996 \cdot 20 = \\ -10.8 \pm 39.2$$

The average reader will have lost £10.8 from 400 roulette spins. Sorry about that. On the other hand, ±39.2 is a broad confidence interval, so some readers will have done rather well. These profit-making gamblers are a sizeable minority, making up 31.2% of people who spin the roulette wheel 400 times. I have noticed this phenomenon if I visit the casino or go to the horse races with a small group of friends. There is usually one person who ends up a winner and does well. It feels like a victory for everyone, especially when he buys the drinks.

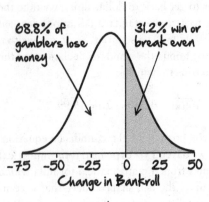

Histogram of player outcomes after putting £1 on a roulette wheel 400 times is distributed according to a bell-shaped Normal curve.

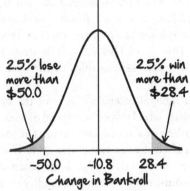

The confidence interval.

95% of gamblers lose less than £50.0 and win no more than £28.4.

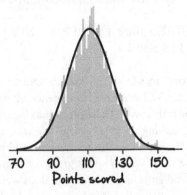

Histogram of points scored per team in every regular season game during the NBA 2018–19 season (shown in grey) compared to Normal curve (solid black line).

Figure 3: The Normal distribution

This is the first key lesson from the confidence equation. The winner might feel that they had a clever strategy, but the reality is that almost a third of people leave a casino as winners. The winners shouldn't be fooled by randomness. They were lucky, not skilful.

*

I have glossed over an important detail: I told you that the distribution of our gambling outcomes follows a Normal curve, but I didn't explain why. The explanation goes all the way back to Abraham de Moivre's presentation in London in 1733.

In his first book on gambling, *The Doctrine of Chances*, published in 1718,[2] de Moivre had worked out the probability of obtaining particular hands in card games and of obtaining winning outcomes in dice games. For example, the probability of having two aces in a five-card hand or the probability of two sixes when throwing a pair of dice.[3] He guided the reader through the calculations involved, setting exercises along the way to increase understanding. This was the type of advice that gamblers would search him out for at Old Slaughter's Coffee House.

In the 1733 presentation de Moivre asked his audience to consider how to calculate the outcome of tossing a fair coin 3,600 times. For 2 coin tosses the probability of getting 2 heads in a row is a straightforward question of multiplying two fractions together: $(\frac{1}{2} \cdot \frac{1}{2} = \frac{1}{4})$. The probability of getting 3 heads in a sequence of 5 coin tosses can be found by first writing down all the possibilities, as follows:

HHHTT, HHTHT, HHTTH, HTHTH, HTHHT,
HTTHH, THTHH, THHTH, THHHT, TTHHH

providing 10 different possibilities. As early as 1653 Blaise Pascal had shown that the number of ways to order k (heads) in a sequence of n (coin tosses) was given by the equation

$$\frac{n!}{(n-k)!k!}$$

The expression $k!$, known as a factorial, is shorthand for writing k multiplied by $k - 1$ multiplied by $k - 2$ all the way down to 1, that is $k \cdot (k - 1) \cdot (k - 2) \ldots 2 \cdot 1$. So, for the example above, where $n = 5$ and $k = 3$

$$\frac{5!}{(5 - 3)! \, 3!} = \frac{5 \cdot 4 \cdot 3 \cdot 2 \cdot 1}{2 \cdot 1 \cdot 3 \cdot 2 \cdot 1} = 10$$

as we also found when we wrote out all of the different possibilities. Since the probability of getting a head or a tail is equal to one half, then the probability of getting k heads from n coin tosses is equal to

$$\frac{n!}{(n - k)! \, k!} \cdot \left(\frac{1}{2}\right)^n$$

For $n = 5$ and $k = 3$ this is

$$\frac{5!}{(5 - 3)! \, 3!} \cdot \left(\frac{1}{2}\right)^5 = 10 \cdot \left(\frac{1}{2 \cdot 2 \cdot 2 \cdot 2 \cdot 2}\right) = \frac{10}{32} = 0.3125$$

There is a 31.25% chance of getting 3 heads from 5 coin tosses.

De Moivre was well aware of this equation, known today as the Binomial distribution, but he also realized how impractical it became when n was large. Solving his $n = 3,600$ coin-tossing problem involved multiplying 2 by itself 3,600 times and calculating $3,600 \cdot 3,599 \ldots 2 \cdot 1$. Try to do it. The calculation is impossible to do by hand and difficult even with a computer.

The trick de Moivre used was to ignore the multiplication itself and study the mathematical form of the Binomial distribution. His friend, Scottish academic James Stirling, had shown him a new formula for approximating large factorials. De Moivre could use Stirling's formula to prove that as n becomes large, the equation above is approximately equal to another equation, namely

$$\frac{1}{\sqrt{2\pi}\sqrt{n/4}} \cdot \exp\left(\frac{(k - n/2)^2}{n/2}\right)$$

At first sight this equation might appear even more complicated than the equation for the Binomial distribution, as it contains square roots, the constant $\pi = 3.141...$ and the exponential function. However, and this was most important to de Moivre's result, it doesn't contain the repeated multiplications found in factorials. We can calculate values for 3,600 or even 1 million coin tosses simply by putting the values of k and n we are interested in into the equation. De Moivre could then use a table of logarithms or a slide rule to solve his problem. Eighteenth-century technology could calculate 1 million coin tosses.

De Moivre built the first ever confidence interval that evening. He showed that the odds of getting either fewer than 1,740 heads or more than 1,860 heads were about 21 to 1, a 95.4% confidence interval.[4]

The equation above is what we call today the Normal curve, and is one of the most important equations in modern statistics. De Moivre didn't appear to realize just how important his equation was, and it wasn't until 1810 that Pierre-Simon, Marquis de Laplace, realized its full potential. Laplace developed a mathematical method called moment-generating functions, which allowed any distribution to be uniquely specified in terms of its average (called the first moment), its variance (called the second moment, about the mean), followed by a series of higher order moments that measure the skew and bumpiness of the distribution. Laplace's moment-generating functions also allowed him to study how the shape of the distribution changes as random outcomes (for example, spins of the roulette wheel and dice throws) are added together. Laplace demonstrated something truly remarkable: irrespective of what is being summed, as the number of outcomes which we sum increases, the moments always become closer and closer to those of the Normal curve.

It took a few years to iron out a few tricky exceptions in Laplace's result (some of which we will come back to in Chapter 6), but by the start of the twentieth century Russian Aleksandr Lyapunov and Finn Jarl Waldemar Lindeberg had tied up the loose ends in Laplace's initial argument. The result that Lindeberg finally proved in 1920 is known today as the Central Limit Theorem or CLT.[5] It says that whenever we add up lots of independent random measurements, each with mean h and standard deviation σ, then the sum of those measurements has a bell-shaped Normal distribution with a mean $h \cdot n$ and a standard deviation of $\sigma = \sqrt{n}$.[6]

To take in the vast scope of this result, consider just a few examples. If we sum the results of 100 dice throws, they are Normally distributed. If we sum the outcome of repeated outcomes of games of dice, cards, roulette wheels or online casinos, they are Normally distributed. The total scores in NBA basketball games are Normally distributed (illustrated in the bottom panel of Figure 3).[7] Crop yields are Normally distributed.[8] Speed of traffic on the highway is Normally distributed. Our heights, our IQs and the outcome of personality tests are Normally distributed.

Whenever a lot of different random factors are added up to produce the final outcome, the Normal distribution can be found. Equation 3 can thus be used to build confidence in any activity that involves repeating the same type of action or making the same type of observation over and over again.

*

In Chapter 1, I showed how a gambler with a 3% edge could turn a £1,000 starting bankroll into £57 million in just one year. By betting and reinvesting, the bankroll grew exponentially. Now I have come to the inevitable catch for my hypothetical gambler. I will call her Lisa. How does Lisa know that she has a 3% edge?

Nate Silver, creator and editor of the political and sport forecasting site FiveThirtyEight, uses the terms 'signal' and 'noise' to explain these situations.[9] In sports betting, the value of the average profit (or loss) from one bet, h in Equation 3, is the signal. If Lisa has a 3% edge, then, on average, each £1 stake will win 3p. The noise per bet is measured by the standard deviation, σ. Like the noise in roulette, the noise in sports betting is much greater than the stake placed. For example, if Lisa stakes £1 on a team at odds of 1/2 then she either loses £1 or wins 50p. We can use the equation on page 58 to show that the standard deviation for this bet is £0.71.[10] The noise, measured by the standard deviation ($\sigma = 0.71$), on a single bet is almost seventeen times larger than the signal ($h = 0.03$). We say that the signal to noise ratio in this case is $h/\sigma = 0.03/0.71 \approx 1/24$.

The casino knows it has an edge because it installed the roulette wheel to give it an edge, a signal to noise ratio of 1/37. Lisa has to rely on past performance to know whether or not she has an edge. This is where the confidence equation is most important in professional gambling. If Lisa has made a profit of h dollars per bet and the standard deviation per bet is σ, then the 95% confidence interval for the estimate of her edge h is found by dividing through Equation 3 by n to get:

$$h \pm \frac{1.96 \cdot \sigma}{\sqrt{n}}$$

For example, if Lisa has placed $n = 100$ bets, and has made an average of 3 cents per bet, then this confidence interval is

$$0.03 \pm \frac{1.96 \cdot 0.71}{\sqrt{100}} = 0.03 \pm 0.14$$

Her edge might be as large as 17p ($0.03 + 0.14 = 0.17$) or it could be that her supposed 'edge' actually loses 11p per bet on average. All edges, winning and losing, between -0.11 and $+0.17$ lie within the 95% interval. The 100 bets she has made have told her very little about whether or not her strategy is working.

While the confidence interval contains zero, Lisa cannot be particularly sure that her signal h is positive and that her gambling strategy works. There is a simple rule of thumb that she can use to establish how many observations she needs to reliably detect a noisy signal. First, let's replace the value 1.96 with 2: the difference between 2 and 1.96 is very small when we are creating a rule of thumb. Now, we rearrange the confidence equation to find the condition under which the confidence interval doesn't contain zero to get[11]

$$\frac{h}{\sigma} > \frac{2}{\sqrt{n}}$$

Taking n observations allows us to detect a signal to noise ratio as larger than $2/\sqrt{n}$.

Below I have tabulated some values to give a feeling for how this rule works.

Number of observations made (n):	16	36	64	100	400	1,600	10,000
Signal to noise ratio detected ($2/\sqrt{n}$):	1/2	1/3	1/4	1/5	1/10	1/20	1/50

Betting and financial edges tend to have a signal to noise ratio close to 1/20 or even 1/50, thus requiring thousands or even tens of thousands of observations to detect. For $h/\sigma = 1/24$, the signal to noise ratio for Lisa's sports gambling, she needs $n > 2,304$ observations. Over 2,000 observations is a lot of football matches. If Lisa thinks she has a 3% edge on the Premier League betting market, she needs six seasons of results to be sure.

During those six years, other gamblers might have picked up on her edge and started to back it. Matthew Benham and Tony Bloom's vast betting operations are always on the lookout for opportunities. Once these two big Bs are in, the bookmakers adjust their odds and the edge disappears. The risk for Lisa is that she doesn't realize that her edge has gone. It takes over 1,000 matches to be confident that an edge exists. It can take just as many expensive losses to realize that it has disappeared. The profits which grew exponentially fast now crash down, and decay exponentially fast.

Most amateur investors are vaguely aware that they need to separate the signal from the noise, but very few of them understand the importance of the square root of n rule that arises from the confidence equation. For example, detecting a signal half as strong requires four times as many observations, and increasing the number of observations from 400 to 1,600 allows you to detect edges that are half as large. It is easy to underestimate the amount of data needed to find the tiny edges in the markets.

*

I called Jan in Berlin, to ask how it was going for him and Marius. It was going really well – so well in fact that Marius had cautioned Jan about what he should say to me. But Jan, as always, wanted to talk numbers. 'What I *can't* tell you, until you speak to Marius again and check it's OK, is that we have had a turnover of £70 million. We made 50,000 bets last month, on an average edge of between 1.5 and 2%.'

The £50 bets we had placed during the World Cup were small change in comparison. When I told Jan that I was now writing about confidence intervals, he referred back to the gambling model we had built together. 'Yeah, well, we made money from that,' he said, 'but to be honest it's not something we would be relying on in the future.' He was right; our World Cup model was built on 283 observations from previous tournaments. Jan had now built a database of 15 billion betting positions made on a whole range of different sports, stretching back over the last nine years.

'We are concentrating on strategies where we have over 10,000 supporting observations,' he told me. This gave them confidence that their strategy had a genuine long-term edge.

Jan and Marius's most profitable edges were based on national differences. Brazilians expect more goals in their games than actually occur. Germans are the opposite: they are pessimistic and always expect a dull 0–0 draw.

'Norwegians are spot on,' Jan laughed, 'perfectly rational Scandinavians.'

I thought back to my conversation during the World Cup with Marius, himself a rational Norwegian, about getting into the head of the gambler. He always saw it as important to have an underlying explanation of any betting strategy. Now he had one: national stereotypes apply.

*

You are looking for a hotel on TripAdvisor. You are happy to stay at a place that gets 4-star reviews or better, but are sceptical about anywhere with 3.5 stars or less. The signal you are looking for here is a half-star difference. Star reviews vary a bit on TripAdvisor. There are always a few enthusiasts who give straight 5s and a few disgruntled individuals who dole out single stars. Overall, however, the noise in reviews is about one star: most reviews are either 3, 4 or 5 stars and the average is slightly over 4.[12]

We can answer the question of how many reviews we need to read to reliably detect a signal to noise ratio of half a star (½) difference using the table on page 66 or we can solve the following equation $\frac{2}{\sqrt{n}} = \frac{1}{2}$, where ½ is the signal to noise ratio. This gives $\sqrt{n} = 4$ or, equivalently, we need to read $n = 16$ TripAdvisor reviews. Instead of looking at the average of all of the hundreds of reviews for a hotel spanning many years, pick out the latest sixteen and take the average. This gives you both up-to-date and reliable information.

And it isn't just hotels that can be rated in stars. Jess isn't sure about her career choice. She has a job at a human-rights organization. It is definitely a worthwhile cause, but her boss is horrible. She rings Jess up all hours of the day and makes unreasonable demands. Her friend Steve has been with Kenny for six months. Their relationship is volatile: one minute it's hot, the next it's cold. The arguments are terrible, but when it works it is wonderful.

The confidence equation offers Jess some guidelines about how many days she should stay in her job and how long Steve should give his relationship with Kenny before he gives up on him. The first thing they need to do is identify the relevant time intervals. Steve and Jess decide to rate every day with 0 to 5 stars. They then plan to meet up regularly to evaluate their respective situations.

On the Friday night of the first week Steve has a massive argument with Kenny, because he refuses to go out with Steve's friends. Steve rings Jess to cry down the phone. He has given three days in his week 1 star each. She reminds him that they agreed not to draw conclusions too quickly. After all, $n = 7$. They can't yet find the signal in the noise. Jess has had an OK week at work, mainly because her pain-in-the-ass boss is away on a trip, so she has collected 3 and 4 stars.

After a month, $n = 30$, they meet up for lunch. They are starting to get a better idea of how things are working out. Steve has had a good few weeks with Kenny. Last weekend the couple went to Brighton for the weekend and, with the inclusion of a few nice dinners, they have had a wonderful time. Steve has rows of 5-star days. For Jess it has been the opposite. When her boss came back she was angry all the time, shouting and losing her patience over the smallest of mistakes. Jess's days are turning into 2, 1 and some 0 stars.

After a little over two months, $n = 64$ and $2/\sqrt{64} = 1/4$. Their level of confidence is now three times higher than that of the first week. For Steve, the good days outweigh the bad days, but there are still small arguments now and again: 3- and 4-star weeks. Jess's boss is a real problem, but Jess has been working on a worthwhile project that she had always wanted to focus on. At best she has a few 3 and 4 stars, but otherwise mostly 1 and 2 stars.

Although each week offers new observations, the square root of n rule means that Jess and Steve aren't gaining information as rapidly as they did when they started meeting. The returns gained by observations diminish. They decide that they are going to set a deadline for their weekly deliberations. After just less than three and a half months (after 100 days), they should be confident enough to decide their futures for good.

It is the big day: $n = 100$ and $2/\sqrt{n} = 1/5$. They look back, not just over the last couple of weeks, but over everything that has happened during that time. For Steve and Kenny the fights have become less frequent. They have started going to cooking lessons together and enjoy nights in making food, often inviting friends over. Life is good. Steve makes a confidence interval. His average star is $h = 4.3$. He calculates the standard deviation in his stars to be $\sigma = 1.0$. The confidence interval for his relationship is 4.3 ± 0.2, a solid average, confidently over 4 stars. Steve decides to stop moaning about Kenny; he is reassured he has found a partner in life.

Things aren't going that well for Jess. Her average star is $h = 2.1$. There have been very few really good days and her standard deviation is lower than Steve's at $\sigma = 0.5$. Her confidence interval is 2.1 ± 0.1. Basically, Jess has a 2-star job. She has already started looking for new positions and on Monday she is going to hand in her notice.

*

In 1964, Malcolm X said, 'No matter how much respect and recognition whites show towards me, as long as that same respect and recognition is not shown towards every one of our people, it doesn't exist for me.'

The idea expressed in these words comes from mathematics. The experience of one person, Malcolm X, or anyone else, provides us with very little information. One person is like one pull of the slot-machine handle. The fact that Jess has a good day at work doesn't tell her about her career in the long term. When people started to listen to Malcolm X, it meant nothing until they listened to African Americans as a group. The struggle of people of colour in the USA against discrimination of all forms, told through the stories of Malcolm X, Martin Luther King and others, was and is the struggle of tens of millions of people.

Joanne hears about a job opening at her work. That evening she meets James at a party and tells him about the opening. James gets excited, saying that it is his dream job, and on Monday he applies. A few weeks later, James has started his new job, and Joanne bumps into Jamal outside the bagel shop. He asks her how it's going at work and Joanne tells him that James has just started a new role. Jamal gets excited, saying that this would be his dream job, and asks Joanne if there are further openings on the way . . .

Joanne is white. So is James. Jamal isn't. Is Joanne racist? No. She would have acted in exactly the same way if she had met Jamal first. It just happened that she bumped into James before she met Jamal.

There is another question, though, about the fact that she met James before Jamal. Simply because James and Joanne occupy the same social group, they meet more often and share information about opportunities. Their mutual support could, indirectly, discriminate against Jamal. He doesn't have access to the same social opportunities as James and Joanne.

Care is required here. We cannot draw this conclusion from Joanne's story. We just have one observation, one anecdote about her interactions with James and Jamal. One event is never enough to build a confidence interval. This is what makes racial discrimination so difficult to pin down. Every individual story is just one observation from which we learn very little. The only way to understand the role of race in a society is to look at many observations and to build a confidence interval.

*

Moa Bursell, researcher and lecturer at the Department of Sociology at Stockholm University, spent two years writing CVs and applying for jobs in Sweden. In total she applied for over 2,000 different positions in computing, as an accountant, in teaching, as a driver and as a nurse. But she wasn't looking for a job. She was testing the biases of the employers she was writing to.

For each application Moa created two separate CVs and cover letters, both detailing similar work experience and qualifications. Once she had finished the applications, she randomly assigned a name to each of the CVs. The first name sounded Swedish, like Jonas Söderström or Sara Andersson; the second sounded non-Swedish, such as Kamal Ahmadi or Fatima Ahmed, to indicate a Muslim Arabic-sounding background, or Mtupu Handule or Wasila Balagwe for a non-Muslim African background. The design of Moa's experiment was based on coin-tossing. If employers are unbiased they would be equally likely to call back the person with the Swedish or foreign-sounding name.

They weren't. For example, in one study of $n = 187$ job applications by Swedish and Arab men, the men with the Arabic names got almost half as many callbacks as the Swedish men.[13] These results cannot be explained by bad luck. We can see this by building a confidence interval. Arab men received 43 callbacks, so the probability of a callback (the signal) is $h = 43/187 = 23\%$. To estimate the variance, we use a value 1 to denote men who received a callback and 0 to denote those who didn't get a callback. Then, in the same way as we did for the spin of the roulette wheel, we calculate the average square distance between these values and h, as well as the average squared distances for the callbacks to Swedish-named men to give $\sigma = 0.649$.[14] If we put these values into Equation 3, we get a 95% confidence interval for callbacks to Arab men of 43 ± 17.3, well below the 79 callbacks received by Swedish men.

It got worse. Moa enhanced the Arab men's CVs, giving them one to three more years of relevant work experience compared to their Swedish counterparts. It didn't help them get a job. In only 26 cases was the more experienced Arab candidate called back, compared to

69 for the less qualified Swedes: again, well outside the confidence interval of 26 ± 15.9.

'What is most powerful about my results', she told me, 'is that they are so easy to understand. There is no arguing with the numbers.'

When Moa lectures on this subject at Stockholm University she can see the students' reaction on their faces. 'When I look at the blue-eyed, blonde students they listen intently. They don't think it is fair, but it doesn't affect them.'

'When I look at those with brown eyes, dark hair and dark skin I see a different reaction. This is about them, and it's about their friends and siblings,' she continues. 'For some it is an experience of finally being recognized. And it can be relief. They can see now that they are not crazy. That their perception of reality is confirmed.'

These students will often talk to her about their own experiences, but others remain silent. 'Hearing about my research can be trau-matic,' she tells me. 'I can see they are upset. They feel it's like being told they are not worth as much and that they don't belong.'

Moa is careful to point out that her study doesn't imply that it is impossible to get a job. The point of the research is to reveal the scale of injustice; it doesn't imply that everyone in Sweden is a racist. It reveals that Kamal Ahmadi and Jonas Söderström have to play for different amounts of time if they are going to win the job lottery.

When a real Kamal Ahmadi applies for a job in Sweden he doesn't know for sure which slot machine he is playing. If he applies for a job and isn't called to interview, he can't claim discrimination. Nor is a real Jonas Söderström able to see the privilege his slot machine gives him. He is qualified for the job; he applied and was called to inter-view. From his perspective, there is nothing wrong with that.

I made this point about Kamal and Jonas to Moa and she said, 'That's true, but you know some people do the experiment them-selves. People with foreign backgrounds have told me that they have asked about a job at a local supermarket and been told that the pos-ition is taken. When they then ask a Swedish friend to phone the shop and ask if the position is still open, they are told that they are wel-come to come for an interview.'

Moa and her colleagues have now sent out over 10,000 CVs to test various hypotheses about the Swedish job market. Some of the findings are depressing. Discrimination against Arab men is strongest in low-skilled occupations. Other findings are possibly more encouraging. Discrimination against Arab women is less pronounced than against Arab men and disappears completely if the women have more work experience.

Studies like Moa's have been repeated across the world with similar results.[15] Moa's study is one example of structural racism, discrimination that is often difficult to spot at an individual level but easy to see through the statistics of the confidence equation. A recent study published in *The Lancet*, the world's leading medical journal, built confidence intervals for measurements of social inequality in the United States, ranging from poverty, unemployment and incarceration to occurrence of diabetes and heart disease.[16] Black Americans differed statistically from whites on every single measure. Toxic waste sites are built close to racially segregated neighbourhoods, governments fail to prevent lead leaking into drinking water, small racial slurs and unintentional insights (telling a black attorney to wait until his attorney arrives), reduced salaries for the same work, targeted marketing of cigarettes and sugar-based products, forced urban renewal and removal, voter restrictions, sub-standard healthcare due to implicit or explicit bias, and exclusion from social networks that could help in finding employment – the list goes on and on. The psychological and physical health of individual African Americans and Native Americans is affected by small amounts of discrimination day after day after day, without anyone necessarily being overtly racist.

Let's go back to Joanne. Is she bumping into more Jameses than Jamals? She decides to use the confidence equation to find out. She thinks about all the people who could be interested in working at the publisher she works for, all the talented people out there, and then she thinks about her own friends, the people she socializes with on a regular basis.[17] Ninety-three out of Joanne's 100 friends are white, compared to 72% of the US population: $93 - 72 = 21$. Her friendships are racially biased. Joanne has checked her privilege. She is woke. She has realized that the people she knows do not represent the population as a whole, and that they belong to a privileged group who share information about job openings in the media. What Joanne should do about this is a difficult question.

Here is what I think. Not a mathematical answer, just what I think. Joanne doesn't need to change her friends. She should be friends with whoever she wants to be friends with. But she does need to consider what she can do about the situation. These are simple things. She can message Jamal, and her other seven minority group friends, when she hears about an opportunity at work or just to check in with them. Jamal has an even more racially biased group of friends than Joanne does: 85 out of 100 are black, in comparison to 12.6% of the US population as a whole and 25% in New York City, where he lives. With one quick message, Joanne has completely changed the demographic of the people who know about a job opportunity.

My view is sometimes referred to as political correctness. I prefer to call it statistical correctness. It is about a statistical awareness that what we experience individually often does not reflect the world at large. It is up to each of us as individuals to work out how statistically correct our lives are and what we should do about it.

*

The confidence equation may have been created for gambling, but it was the natural sciences and, eventually, the social sciences that it transformed. The first member of TEN to realize the true scientific power of the Normal curve was Carl Friedrich Gauss, who used it in 1809 to describe errors in his estimate of the position of the dwarf planet Ceres. Today, the Normal curve is often referred to as the Gaussian, a somewhat unfair attribution given that the equation is clearly stated in the second edition (1738) of de Moivre's *The Doctrine of Chances*.[18]

Statistics became fully integrated into science through the massive advances during the nineteenth and start of the twentieth century. After the Second World War, confidence intervals became an essential part of scientific writing, forcing researchers to show confidence that their conclusions weren't just a result of random chance. The most recent scientific paper I submitted contained more than fifty different confidence interval calculations. The existence of the Higgs boson was only confirmed when the statistics reached a 5-sigma confidence level, which implied that the probability that the results of the experiment would have occurred if there was no Higgs boson was 1 in 3.5 million.

TEN's progress in the social sciences was initially slower than in the natural sciences. Until recently, a caricature of a sociology department might have described the staff as shabbily dressed men who worshipped dead German thinkers and women with purple dyed hair, brought in during the 1970s to shake things up with postmodernist ideas. They argued and discussed but could never agree. They created definitions and frameworks for thinking and argued some more. Those on the outside had no idea what they are talking about.

Up until the turn of the millennium, this caricature contained a large element of truth. Statistics and quantitative methods were used, but sociological theory and ideological discussion was seen as the way to study society. In the space of a few quick years, TEN blew this old world away. Suddenly researchers could measure our social connections using Facebook and Instagram accounts. They could download every opinion blog ever written and understand our methods of communication. They could use government databases to identify the factors that led us to move between jobs and neighbourhoods. The structure of our society was laid wide open by a new availability of data and statistical tests, establishing a level of confidence in every result.

Ideological arguments and theoretical discourse were pushed to the outer perimeters of the social sciences. A theory was worth nothing without data to back it up. Some of the old guard of sociologists joined the data revolution, others were left behind, but no one working at a university could deny that the social sciences had changed for ever.

*

The transformation of social science by the use of data has not been noticed by everyone. I sometimes read an online magazine called *Quillette*. This magazine prides itself on continuing a tradition of public scientific dialogue that can be traced back to Richard Dawkins in the 1980s and 1990s. Its stated aim is to give a platform for free thought, even dangerous ideas, by which it means that it is happy to publish views about gender, race and IQ that are not necessarily 'politically correct'.

Quillette articles regularly attack research in social science. One of their favourite targets is identity politics. I recently ended up reading one article, written by a retired psychology professor who claimed that social science was deteriorating into 'incoherence and nonsense'. He took issue with a book edited by Tukufu Zuberi and Eduardo Bonilla-Silva, *White Logic, White Methods: Racism and Methodology*, in the social sciences.[19] The book investigated the degree to which the methods used by social scientists are determined by 'white' culture. Based on his doubts about the 'white methods' claim, the professor went on to counter-claim that he couldn't find any evidence of systematic racism anywhere in society. He suggested that the 'abilities and interests of African Americans' might offer a better explanation for the differences we observe.[20]

The authors of many other *Quillette* articles, rather than reviewing the data, tend to try to fire up a debate with academic sociologists and left-wing activists. There is less focus on numbers and more on a culture war of ideas. As I'll show in Chapter 7, there is very little inherent difference between biological races (in fact there is no such thing as biological race), while there is, as reported in *The Lancet* article referenced above, rather a lot of evidence for structural racism in the USA.

I emailed a copy of *The Lancet* article to the author of the *Quillette* article and suggested he review its contents. We had a few friendly emails back and forth. It turned out that, when it came to the study of animal behaviour, we had quite a few research interests in common.

Then, a few weeks later, he sent me his new magnum opus, an attack on the very idea of structural racism. He claimed, among other things, that it was always impossible to prove racism, because so many other factors would have to be ruled out. The retired professor seemed to miss the whole point of doing statistics, that is, to detect patterns of discrimination over lots of repeated observations. He reiterated his calls for more race biology.

In response to an American version of Moa Bursell's CV study, demonstrating discrimination against African American names, the retired professor wrote, 'Is this racism? We don't know the hirer's previous experience. Perhaps she had had a bad experience with black hires in the past?'

Is this racism on his part? Well, yes, it is. No confidence interval required on that one. And, to my utter amazement, a few months later *Quillette* published his ill-informed musings. Thankfully, the 'bad experience with black hires' phrase had been removed. But the published article continued in the same tone, denying without evidence the basic facts in *The Lancet* article.

Quillette is not alone in adopting this approach to the social sciences. In the UK, the publication *Spiked*, the online reincarnation of the 1990s print magazine *Living Marxism*, regularly attacks gender politics and the idea that structural racism exists. The 'Culture War' thread on social media site Reddit allows anyone and everyone to join the debate. The same concepts permeate the 'Intellectual Dark Web', a self-proclaimed movement for free ideas, expressed on YouTube and in podcasts, that demands the right for all ideas to be heard. The Intellectual Dark Webbers don't just write about gender and race, but their focus on challenging political correctness means that it usually isn't long before the discussion moves on to these two 'taboos', as they like to refer to them.

The king of the Intellectual Dark Web is Jordan Peterson. Like *Quillette*, he is waging a war against what he perceives as the political correctness that has taken over the social sciences. He believes that a leftist ideology has driven academics to focus on questions of gender and racial identity. He describes universities as a place where there is a fear of saying the wrong thing. Ultimately, he argues, this has a negative effect on society at large. White people are unfairly attacked for their privilege and women are potentially given unfair advantages in hiring decisions.

Last time I flew business class (which I am sometimes forced to do) the two tech guys sitting behind me spent the whole trip discussing how well-dressed Peterson was and how he handled himself in an argument. I wanted to turn around and protest, but couldn't actually pin down exactly what was wrong with what they were saying. He is well-dressed, he can argue and even cries at the right times when interviewed.

I read Peterson's book, 12 *Rules for Life: An Antidote to Chaos*.[21] I enjoyed it. It is full of interesting anecdotes from his life. Some good tips on how to be a man. Good title too. But it isn't modern social science. Nowhere close. It is a privileged white man spinning his own personal casino wheel and telling us how lucky he is.

Modern academia is very different from the way Peterson portrays it. I work with lots of social scientists and don't think I have ever met a single one who is scared to say what they think. Quite the contrary. They won't shut up. Having controversial ideas, thinking about different models is an important part of our job.

Where modern scientists are constrained is by the confidence equation. If we want to test our model, we have to work hard to collect data. Social science is no longer about anecdotes or abstract theories. It is about creating and sending out thousands of CVs, or about carefully reviewing the literature to identify pathways to structural racism. It is about hard work, not looking good in a suit or appearing to think hard before you answer a question.

Moa Bursell developed a left-wing political view from her early teens. 'Many of my best friends at that time [the early 1990s] had a foreign background,' she told me. 'When we would go out in the evening, they were scared and many of them were harassed by neo-Nazis. I would have to run with them from trouble. These experiences led me to politics.'

When Moa talks about her formative years, she is open and emotional, in contrast to the dispassionate way she talks about her scientific results. She also told me about the time, many years later, when a group of immigrant teenagers visited her university, brought there by a young equality activist. The activist wanted Moa to tell the teenagers about her research on job searches, but Moa was reluctant, scared that it would be perceived the wrong way. She had been right to be cautious. When Moa explained her results, the response was one of anger. 'If we have no future, why should we bother going to school, then?' the kids asked.

Moa was deeply shocked by the experience and disappointed with herself. 'I know that many immigrants feel they don't belong from the time they enter school,' she said, 'and it was like we brought them to the university just to tell them that they would be discriminated against in the workplace too.' Simply communicating the problem to those affected was not always part of the solution.

Like all social scientists, Moa has ideals, dreams and political opinions. These are her models of the world. There is nothing un-scientific about finding motivation in our beliefs and experiences, as long as the models are then put to the test against the data. As Moa put it when I asked her how she started her research career, 'I believe, as I think [sociologist] Max Weber said, that you should choose your research topic with your heart, but then approach it as objectively as possible.'

She went on, 'The CV experiments interested me because you can't argue with the results. I was studying real people, not in a lab. Every-thing is controlled in the experiment and the results are simple and easy to understand.' The model is tested against data. Moa told me that she was surprised when she found no gender discrimination in how employers assessed CVs, or even in gender compensation: there were more callbacks to women in professions like computing when they were under-represented. It went against her personal preconcep-tions. 'But there it is,' she said, 'I can't argue with it now either.'

One of the issues that Jordan Peterson is best known for debating is the gender pay gap. He has correctly pointed out that the fact that women in the USA are paid 77¢ on average to every dollar paid to a man is not in itself evidence of discrimination. He makes a useful distinction between equality of outcome and opportunity.[22] Women are paid less than men partly because they work in jobs that typically pay less, such as nursing. It is plausible that women might have had opportunities to work in better paying jobs, but have chosen different career paths from men. Peterson also argues that women might be biologically less suited to certain types of better paying work. In short, he says, we can't use pay gaps themselves to argue for discrimi-nation; we need to test whether or not they had the same opportunities as men.

Equality of opportunity is exactly what Moa tests with her CV experiments. When Muslims apply for jobs, they have lower callback rates than native Swedes and are thus subject to discrimination in opportunity. Likewise, Moa's results reveal equality in opportunity for native Swedish women at the point they send in CVs. In this particular case, Peterson's claim that there is no discrimination in opportunity is correct.

However, while equality in outcome can be measured using a single number (the gender pay gap, for example), equality in opportunity cannot. There are many ways in which women can be prevented from achieving their full potential and thus many potential barriers to opportunity need to be investigated.

Luckily, social scientists are working hard to identify these barriers. In 2017 Katrin Auspurg and her colleagues interviewed 1,600 German residents, presenting them with a series of short descriptions of the age, gender, length of service and role at work for a hypothetical person, before asking whether the specified salary was fair.[23] The respondents tended to rank women as overpaid and men as underpaid. On average, the respondents, both male and female, believed that the women in the scenarios should be paid 92¢ for every dollar paid to a man for doing the same job. At the same time, the vast majority of the respondents, when asked the question directly, agreed that men and women should be paid the same. There is a big difference between what we say and how we act in practice. The respondents in this study didn't even realize that they actually recommended that women should be paid less than men for the same job.

Assessment by scientists of CVs sent for lab assistant jobs in the US are biased against women, it was found in a study of 2012.[24] Again, both male and female scientists consider females' CVs to show them as being less competent. Women who have only male maths professors instructing them are also less likely to continue with the subject than if they have female professors.[25] In an experiment with high school students, girls working in classrooms containing stereotypical geeky objects (*Star Wars* items, tech magazines, video games, science fiction books, etc.) are much less likely to express an interest in continuing in the subject than those working in a non-stereotypical environment (with nature and art pictures, pens, a coffee maker, general magazines, etc.).[26] In high schools in Canada, girls tend to rank themselves as weaker in mathematics than boys, despite performing at the same level in exams.[27] In a work negotiation experiment, conducted on students at an American university, women were found to be just as effective as men when negotiating on someone else's behalf, but less effective when negotiating on their own behalf. These differences are explained by a fear of backlash if they win the argument, a fear not experienced to the same degree by men.[28]

These are just a few of a large number of studies recently reviewed by Sapna Cheryan and colleagues which reveal the barriers to opportunity based on gender.[29] Women and girls find it more difficult to express themselves freely; they fear retribution; they are devalued by both men and other women; they are presented with fewer role models; they undervalue themselves; and they are implicitly discriminated against when they apply for certain jobs: this is the statistically correct way to view the schools and workplaces we go to every day. Since most people, including Jordan Peterson, agree that we should strive after equality in opportunity, the answer is simple: we need to educate people about the research results that have identified the biases within our society.

Peterson, bizarrely, draws the opposite conclusion. He attacks academic research on questions of diversity and gender, claiming they have a left-wing agenda and that studies are run by Marxists. Here he is just plain wrong. Social scientists – like Moa Bursell, like Katrin Auspurg and like Sapna Cheryan – are deliberately studying opportunity instead of outcome. These researchers may well be motivated by a desire to create a fair playing field for everyone, but that desire for fairness makes it even more essential that they are not swayed by any political views they might (or might not) hold. In all the research I mention above, and in many more examples, the aim is to find out where equality in opportunity is lacking so that the problem can be solved. There is no evidence of ideological bias on the part of the researchers.

Peterson never mentions these studies. Instead, he focuses on psychological differences between men and women. In an interview in January 2018 with Cathy Newman on *Channel 4 News* in the UK, which later went viral on YouTube, he argued, 'Agreeable people are compassionate and polite, and agreeable people get paid less than disagreeable people for the same job. Women are more agreeable than men.'[30]

There are several good reasons why psychological explanations like this are not as convincing as confidence interval testing of specific models. The rationale behind posing direct and context-relevant questions such as 'How would you rank this CV?' or by observing how women and men negotiate is that by understanding individuals' actions we can provide causal explanations of how inequality arises.[31] In contrast, agreeableness is established through a self-reported personality test, where people answer general questions, such as 'I sympathize with others' feelings'. Saying someone is 'agreeable' is just a way of summarizing answers from these questionnaires. It isn't immediately obvious why being agreeable would be a hindrance to a higher salary. It could go either way. Maybe nice people are rewarded for their friendliness or maybe they are poor negotiators? Agreeableness could have different effects depending on the careers, the skills involved in the job and the seniority of the people involved.

Personality tests don't in themselves provide an explanation, thus in order to relate agreeableness to salary an extra test is required. One study of fifty-nine recent graduates in the USA found that in early career situations, agreeable people do receive lower salaries.[32] But it also found that women were paid significantly less than men. Even when accounting for all other personality traits, general mental ability, emotional intelligence and success in the job, agreeableness was the only factor which helped explain the pay gap, and then only to a small degree. Disagreeable women were still paid less than disagreeable men and agreeable men were paid more than agreeable women. In fact, what this study showed, in contrast to what Peterson told Newman in the interview on Channel 4, is that non-gender-related factors completely failed to explain the pay gap.

The lack of a clearly stated model of how personality influences salary makes it very difficult to talk in concrete terms about how personality affects opportunity. Even if we did finally establish that there is discrimination against agreeable people in terms of pay, rather than direct discrimination against women, we would still need to answer the question as to whether or not that was fair? Some explanations – such as agreeable people not representing their company's best interests – might be considered fair, while others – such as bosses are taking advantage of agreeable employees by paying them less – might not. Talking in general terms about personality doesn't help us understand the real issues.

It is also important to look more closely at what Peterson actually means when he says that women are more agreeable than men. Here we can apply the confidence equation. Psychologists have now conducted personality studies on hundreds of thousands of people. As we saw earlier in the chapter, the more observations we have the better we can detect a signal hidden in noise. For example, if we conduct $n = 400$ personality surveys, then we can detect a difference even when the signal to noise ratio is 1/10. With lots of data we can use the confidence equation to identify even very small differences between the agreeableness of men and of women. And small differences are exactly what we find between men and women when it comes to personality. The signal to noise ratio for agreeableness, the personality trait that varies the most between men and women, is around 1/3. There is one unit of signal for every three units of noise.

To understand how weak this signal is, imagine picking one man and one woman at random from the population. The probability that the woman is the more agreeable is just 63%. Think about what this means in practice. You are standing behind a closed door and are about to be introduced to Jane and Jack. Is it reasonable for you to walk into the room and say 'Jane, I think you are going to agree with me more than Jack because you are a woman', before turning to Jack and saying 'Right, you and I are probably going to have an argument'?

No. That would be statistically incorrect. There is a 37% chance you have made a mistake.[33]

Peterson claims, as he did for example on the Scandinavian talk show *Skavlan*, that psychologists have 'perfected, at least to some degree, the measure of personality with advanced statistical models'.[34] Then he mentions the vast number of personality questionnaires that have been carried out on hundreds of thousands of people. After this, he correctly concedes that men and women are more similar than they are different. Then he asks us to consider 'where are the largest differences?' before going on to tell us that 'men are less agreeable ... and women are higher in negative emotion, or neuroticism'.[35]

While not entirely wrong, his argument is somewhat disingenuous. The implication is that scientists have now pinpointed large gender differences of a very specific sort, using vast quantities of data. The proper interpretation is that out of the hundreds of different ways that researchers have thought up to ask men and women about how they view themselves, and the hundreds of thousands of people surveyed, almost none of these studies have revealed any really strong gender differences in personality. In fact, the most remarkable result from the last thirty years of personality research is the gender similarity hypothesis. This hypothesis, first proposed by Janet Hyde, Professor of Psychology and Gender and Women's Studies at the University of Wisconsin-Madison, in 2005 in the journal *American Psychologist*, doesn't say that men and women are the same. She says that there are very few statistical differences between personalities that depend on gender. Hyde reviewed 124 different tests of personality differences, and found that 78% of these tests revealed negligible or small differences between the genders (signal to noise ratios smaller than 0.35).[36] The hypothesis has stood the test of time: ten years later a new, independent review found only 15% of 386 tests provided a signal to noise ratio greater than 0.35 between gender and personality.[37]

In terms of neuroticism, extroversion, openness, positive emotions, sadness, anger and many other personality traits, men and women differ only slightly or not at all. In a more recent review, Hyde found that gender differences are also small in mathematics performance, verbal skills, conscientiousness, reward sensitivity, relational aggression, tentative speech, attitudes to masturbation and extramarital affairs, leadership effectiveness, self-esteem and academic self-concept.[38] The sexes differ most in interests in things versus people, physical aggression, pornography use and attitudes about casual sex. It turns out that a few of our preconceptions are true after all. Our individual personalities vary a lot. Men are very different from each other. Women are very different from each other. But, in general, it is statistically incorrect to say that men and women's personalities are very different.

The danger is in Jordan Peterson's claims that research on gender has somehow been subverted by 'left-wing' and 'Marxist' forces. The opposite is true. Janet Hyde, whose undergraduate degree was in mathematics, is part of the statistical revolution measuring equality in opportunity that has swept ideological thinking out of psychology and the social sciences. She has received multiple awards for her research, including three different awards from the American Psychological Association, the largest scientific and professional organization for psychology in the USA. Gender differences have been studied meticulously, to the point that every small difference has been documented. Similar results have been found studying differences between male and female brains: the variation between human brains of any gender in structure and function is much larger than the variance between male and female brains.[39] The irony is that it is only because of this vast body of rigorous research that Peterson can cherry-pick results to support his own, ideologically motivated standpoint.

The confidence equation teaches us to replace anecdotes with observations. Never rely on one person's story, even your own. When you are winning, think carefully about whether or not your streak has lasted long enough for you to really put it down to skill. There is always someone who is lucky, and maybe this time it is you. Search out other stories and collect statistics. As you make more observations, think in terms of the square root of n rule: in order to detect a signal half as strong you need four times as many observations. If you really are 'up', that is statistically better off than those around you, then use the confidence interval to check your privilege. Be statistically correct: understand the advantages and disadvantages you have in life. Become confident, not by fooling yourself, but by understanding exactly how society shapes your life. Only then can you find and claim your edge.

4

The Skill Equation

$$P(S_{t+1}|S_t) = P(S_{t+1}|S_t, S_{t-1}, S_{t-2}, \dots, S_1)$$

I am sitting in a café in the late afternoon, and watch him come in. He shakes one of the waiter's hands and then does the same thing with the barista, exchanges smiles and a few words. He doesn't see me at first, and as I stand up to go over to him he spots someone else he knows. A round of hugging ensues. I sit back down again, waiting for him to finish.

His celebrity here partly derives from his former life as a professional football player, and because his face is often on TV, but he is also popular because of how he holds himself: his confidence, his friendliness, the way he takes the time to talk to people – sharing a few words with everyone.

Within a few minutes of sitting down with me, he is into his spiel. 'I think I make a difference because I show them my way of doing things. I think that's lost sometimes,' he says. 'I just do my thing, I tell it as it is and I am honest, because that's what is needed in this game.

'I've got a lot of contacts. A lot of meetings like this one, you know, keeping connected. You see, people want to talk to me because I have a unique way of seeing it. Because of my background, you know, a way that no one else has quite got, and that's what I'm aiming to deliver when I sit down with you . . .' These observations are interspersed with anecdotes of his playing days, a bit of name-dropping and rehearsed stories, complete with well-timed jokes.

He smiles, looks me straight in the eyes and, at times, makes me feel like I've asked for all this information. But I haven't asked for it.

I wanted to talk about using data, both as it is employed in the media and within the game of football. Unfortunately I'm not getting anything useful.

I call this type of man Mr 'My Way', after the song that Frank Sinatra made famous. The careful steps, the standing tall and the seeing it through provide the basis for each of his stories. It can make a beautiful melody, and for the two or three minutes during which my current Mr 'My Way' is hugging and greeting his way into the café, it entertains those he meets.

But it only works provided he moves from one person to the next. Now, here am I, stuck in this position, with nowhere to go.

The first few times I spoke to a footballer Mr 'My Way', I am ashamed to admit, I bought their stories. Since the publication of my book *Soccermatics* in 2016, I have been invited to visit some of the world's leading football clubs and had their representatives visit me. I've been invited to appear on radio and TV to talk about the game together with former professionals. It is intoxicating to move from an academic environment to mixing with ex-footballers, TV personalities, scouts and board members of Premier League football clubs. I've enjoyed hearing behind-the-scenes stories about players and big matches, and finding out about life at the training ground. Moving from being a fan to being someone who is confided in by those close to the action was, to use the biggest footballing cliché possible, a dream come true.

I still love hearing those stories and seeing the real world of football for myself. But more often than not, the interesting bits are accompanied by 'heroic' tales of the 'vision' of Mr 'My Way', followed by accounts of how their progress has been foiled by a cheating adversary or how they could do things better than anyone else if they had been given half a chance.

Because of my background in maths, these guys often feel they have to explain their thinking process to me. They start by telling me that I have a different way of looking at things than they do, without actually asking me how I look at things.

'I think stats are great for thinking about the past,' they will tell me, 'but what I bring is insight into the future.'

After that he will explain how he has a unique ability to spot a competitive advantage. Or how it is his self-confidence and strong character that helps him to make good decisions. Or how he has cracked a way of picking out patterns in data that I have (he assumes) missed. His tales tend to include a digression to times that didn't go quite as well for them. 'It was only when I lost concentration that I started to make mistakes,' he tells me. But he always returns to emphasizing his strengths: 'When I stay clear and focused I get it right.'

What I hadn't understood when I started working in the football industry was just how much time I would have to sit listening to men telling me why they believed they were the special one.

I should have known better, because this doesn't just happen in football. I have experienced the same thing in industry and business: investment bankers telling me about their unique skill-sets. They don't need maths, because they have a feeling for their work that their quantitative traders (known as 'quants') can never have. Or tech leaders explaining to me that their start-up succeeded because of their unique insights and talents. Even academics do it. Failed researchers describe how their ideas were stolen by others or, when they succeed, they tell me how they stuck to their principles. Each of them did it their way.

Here is a difficult question to answer. How do I know whether someone is telling me something useful or not?

The guy I'm sitting with now is obviously full of it. He has talked about himself non-stop for the last hour and a half. But many other people do have something useful to say, including, on occasion, Mr 'My Way'. The question is how to separate the useful stuff from the self-indulgent stuff.

*

The applied mathematicians' approach to this question is to divide everything that people say to you into three categories. The first two categories were discussed in previous chapters: model and data. Models are our hypotheses about the world and data is the experiences we have which allow us to establish the truth, or otherwise, of our hypotheses. The Mr 'My Way' I am talking to now is producing a lot of a third category: non-sense. He is telling stories about his own triumphs, failures and feelings without revealing anything concrete about how he thinks or what he knows.

I use the hyphen in 'non-sense' in order to make you think a bit more about the word. This is a trick borrowed from Oxford philosopher A. J. Ayer, who inspired my own way of seeing mathematics. Ayer recognized that 'nonsense' is a very provocative word, but he used it to describe information that doesn't come from our senses. How Mr 'My Way' feels, how he perceives his successes and his failures is not based on observations or things we can measure. Ayer proposed that when Mr 'My Way' or anyone else tells you anything, you should ask whether or not that statement is verifiable. Can you, in principle, check or verify whether a statement is true or not using data you get from your senses?

Verifiable statements include things like 'the plane we are on is about to crash', 'Rachel is a bitch', 'miracles occur', 'Jan and Marius have an edge on the betting markets', 'Swedish employers make racially biased decisions about who to invite to interview', 'Jess should quit her job if she wants to be happier' and so on. These are exactly the sort of statements I have formulated as models in this book. When we compare models to data we are verifying the extent to which they are true.

We don't demand that we access the data to check the verifiability of our models. When Ayer published his book *Language, Truth and Logic*, in 1936, which explained the principle of verifiability, there were no pictures of the far side of the moon. So the hypothesis about the existence of mountains on the moon's far side could not be proclaimed as true or false. It was verifiable, though, because it could be checked in principle, and when the Soviet *Luna 3* spacecraft flew around the moon in 1959 the facts were checked.

Statements about the feelings of Mr 'My Way' and his sense of self-belief are different. His collection of stories might contain titbits of information, names of real people and events that actually occurred, but they are not verifiable. We can't set up a test to confirm whether or not he has a 'unique way of seeing it' or exactly what he has got 'that no one else has quite got' or how he knows 'what is relevant and what isn't'. Such a test is impossible because he can't properly explain the basis for these statements. He can't separate his feeling from the facts and we can't reformulate his statements as a model that can be tested against the data. Instead, what defines the song of Mr 'My Way' is the mishmash of personal ideas. What he says is neither data nor model. It is, quite literally, senseless.

*

La Masia, Barcelona. When it comes to a deep intellectual approach to the beautiful game there is nowhere that can be compared to Barcelona football club's training facilities. Founded in 1979 by footballing legend Johan Cruyff as an academy for young players, it has developed a philosophy that runs through everything done within the club.

I walked past a small group of fans, each of them hoping to catch a glimpse of the players as they went in and out of the front gate, and found the side entrance of the new La Masia. Just as many universities have relocated from older, traditional buildings to shiny new structures, so too has Barcelona's academy and sports research institute moved from its original location in a farmhouse to a modern, glass-fronted block.

I had been invited to La Masia by Javier Fernandez de la Rosa, head of sports analytics at Barcelona FC and PhD student in Artificial Intelligence. He had asked me to give a presentation of my recent work and talk to them about ways of analysing the game.

The inside of the new La Masia is also like a modern university department in that it houses both education/training and research. The first team players had just finished their training session and the youngsters were busy on another pitch. Javier sat in a brightly lit office with an array of monitors in front of him and rows of books behind him. At other clubs I have visited the facilities have been dominated by the training, and analysts could be found crammed into out-of-the-way areas. Here the players had everything they could ever need and the researchers had their own space to work and reflect – to plan and improve the team's style of play. The organization of space within La Masia reflected the game of football as I saw it now: the mind working together with the body.

Javier and I set to work immediately. We went into his office, sat down with both our computers open and started to compare notes. How do you evaluate passes? How do you track player movements? How do you segment a match into different game states? What is your definition of a counter-attack? How do you model pitch control? The questions and answers shot back and forth. Data, model, data, model, data, model and then some more model. It went on and on.

At some point – quite suddenly it seemed to me – Javier said it was time for me to give a seminar to the rest of his group. We went to a spacious seminar room, plugged in my laptop to a large screen and I stood up and started again, this time presenting to an audience of coaches, scouts and analysts. Then a group of five or six in the front row started to interrupt to ask about the data I was using, my assumptions and my results. They told me about their own findings and how I could do things better.

The Barcelona sports analytics team delivered exactly what I love about research: a deep dive straight into model and data. A perfect day of research was topped off with front-row seats to see Lionel Messi and company in action in the evening. As the sun went down over Camp Nou, I was as close as I will probably ever get to the same moving body I had, earlier that day, viewed as a curve on the coordinates of my computer screen.

*

I focused my presentation in Barcelona on one player in particular. At the time, a few months after the 2018 World Cup, I was very interested in Paul Pogba. And so, too, if newspaper rumours at the time were to be believed, was Barcelona.

I have been a fan of Pogba for a long time because, more than any other player at the top of his game, he defines the teams he plays for. While Lionel Messi is Barcelona's talisman, the philosophy of the club he plays for is to become more than the sum of its parts, rather than focusing on any one individual. Cristiano Ronaldo certainly has presence on the pitch, but ultimately he is a traditional, very athletic striker. The style of football at Juventus or Real Madrid is not built solely around his abilities.

When Paul Pogba plays for Manchester United he *is* that team and, during the World Cup, he defined his nation, France, who won the trophy. That is my hypothesis, but how can I verify it? Unlike Messi and Ronaldo, Paul Pogba doesn't score a vast number of goals. In the World Cup he scored only once, in the final, which is itself a merit, but lots of other players scored more times than he did. So goals alone don't explain his skill.

The mathematical idea I used to evaluate Pogba was to focus on the contributions he makes towards his team scoring a goal, rather than the goals he scores himself. At this point, a football fan might ask if I am talking about assists, the passes made by players that lead to a goal. A first assist is the pass made to the player who scores; a second assist is the pass to the player who passed to the goal scorer, and so on. Counting assists is part of the approach I'll take, but still only a very small part. Instead of giving events like goals or assists special status, I evaluate every action that happens on the pitch: tackles, passes, interceptions and so on. My aim is to measure how each of these actions increases his team's probability of scoring and decreases his opponent's probability of scoring.

To achieve this aim, first we need to think about how to describe a football match in numbers. Imagine a pass made from a position (x_1, y_1) on the pitch to another position (x_2, y_2). To envisage these pass coordinates, take a bird's-eye view of a football pitch. The x-direction runs along the touchline and the y-direction runs along the goal line. The coordinate $(0,0)$ is the corner flag on the right of the attacking team's goal line. The coordinate $(105,68)$ is the corner flag on the opposite side of the pitch (a typical professional football pitch is 105 metres long and 68 metres wide). Every pass during a match can be described in this way: $(10,30) \rightarrow (60,60)$ is a long kick by the goalkeeper out to the wing; $(60,60) \rightarrow (60,34)$ takes the ball in to the centre of the pitch; and $(60,34) \rightarrow (90,40)$ moves the ball into the opponent's box. Imagine a football match as a sequence of coordinates updated by the passes and dribbles performed by the players. Every sequence of play, or possession chain as we call it, can be broken down into a description of actions occurring at x and y coordinates on the pitch.

What we now want to do is determine how each player's individual actions in these possession chains increase their team's chance of scoring and/or decrease their opponent's chance of scoring. To do this, I am going to make a mathematical assumption. As a general rule, when a mathematician tells you that she is going to 'make an assumption', what she means is that she is now going to say something untrue and ask you to suspend your disbelief and use your imagination instead. This is a bit different to our everyday use of the word. For example, I might say to my wife about guests we have invited to dinner, 'I assume they will come around 7 p.m.' Or I might say, 'I assume we are going to lose again', when my team is two goals down with five minutes left to play. These are both things that are likely to be true, but they are not mathematical assumptions.

In maths, we use the word 'assume' to describe a set of things that we know are not necessarily true, but that we don't want to worry about for now. What I want, please, is for you to suspend your disbelief and then together we will look at where this assumption takes us, without discussing the assumption itself any further. But it is important that the assumption is stated from the start, because it is the basis of our model, and when we compare models to reality we need to be honest about their limitations.

The assumption I am making here is that the quality of a pass in football depends on its start and end coordinates and not what comes before and after the pass, or which players are on the pitch when the pass is made, or anything else like that. So, if Pogba can make a pass from the middle of the pitch, say at coordinate (60,34), into the penalty area at coordinate (90,40), then this pass will always, irrespective of everything else happening in the match, have the same effect on France's chance of scoring.

But the assumption is clearly incorrect. For example, in the World Cup match against Peru, within the space of a single minute, Pogba made two passes into the penalty area from around about the same point on the pitch. The first pass was chipped over the defence to reach M'bappé who, despite some acrobatics, failed to divert the ball past the goalkeeper. The second pass went along the ground to Olivier Giroud, whose shot was initially blocked by a defender, before ending up in the path of M'bappé, who this time scored France's opening goal. My assumption is that these two passes – one that led to a missed chance and one that led to a goal – both had about the same value to France as a team.

By suspending our disbelief, we can build a model of everything that happens in a football match. Together with my colleague Emri Dolev, I used a database of the start and end coordinates of every pass made over many seasons of top-level football in the Premier League, the Champions League, La Liga, the World Cup, etc. We looked at every single pass to see whether or not it ultimately led to a shot on goal. This allowed us to fit a statistical model linking the start and end coordinates of a pass to the probability of a goal being scored (see Figure 4). In this way, we could assign a value to every pass, independent of what came before or after it.

Once Emri and I had assigned a value to every action in every match, then we could finally evaluate Paul Pogba. He stands out for two reasons: his ability to get the ball back in midfield and his ability to immediately turn defence into attack with long, accurate passes. He played some amazing passes during the World Cup, taking the ball back from near the middle of the pitch, turning around and delivering it to the feet of a teammate deep within the opponent's half. He increased France's chance of scoring more than any other player in the team.

Barcelona already has a player who plays a similar role to this: Sergio Busquets. While Lionel Messi is the star attacking player at Barcelona whose name everyone recognizes, Busquets is the engine which drives the team, starting attacks from midfield. Busquets and Pogba differ in many ways, but in their ability to impose themselves in midfield they are very similar. Busquets is five years older than Pogba; as engines get older, they become less effective over time.

The model Emri and I developed can be applied to every professional player in every match. It can evaluate them in the same way it evaluated Pogba, within seconds. This allows teams to pick out players that fill their exact requirements. When a player leaves, they can find replacements who are made to measure.

The traditional way of evaluating performance is to have scouts watch matches and write scouting reports. A technical director for one leading club recently showed me his database of potential recruits. He could look up seventeen-year-olds playing Division 3 football in

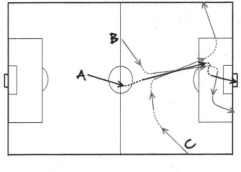

Each chain of possession is assigned value 1 if it ends in a goal, 0 if it goes out of play. So, chain A has value 1 and chains B and C have value 0.

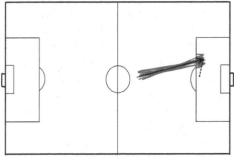

The Markov assumption implies that the value of a pass is the proportion of times it occurs in a chain which leads to a goal. In this case, out of 10 similar passes 1 leads to a goal, i.e., 0.1.

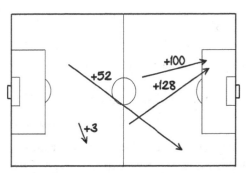

We multiply the proportion by 1,000 to assign points between 0 and 1,000 for every pass Pogba makes. The sum of all points is a measure of Pogba's skill in passing the ball.

Figure 4: How the Markov assumption can evaluate passes in football

Sweden or fifteen-year-olds in the Brazilian junior leagues. A green tick next to these players indicated that a scout had been to watch them play. The director could click on the player and read multiple reports by different scouts on any player in the world.

Our model is complementary to this approach. It looks specifically at the player's ability to move the ball from one set of pitch coordinates to another. When a scout is evaluating a player, he is using his experience to assess the player's positioning on the pitch, his awareness of those around him and how he interacts with teammates. No scout, however good, can claim to be assessing every pass a player has ever made in the Premier League. But a model can make this claim.

When I talk to football scouts and coaches, this is exactly how I describe my assumptions. Instead of saying things like, 'Statistics show that Pogba was the best midfielder at the World Cup', I say, 'If we are interested in how far a player progresses the ball from midfield, then, both at the World Cup and when playing for Manchester United, Pogba is one of the top-ranked players in the world.'

Setting out our assumptions, as well as our conclusions, when we talk to others is essential – not just when we are discussing football, but when we are discussing anything that we care about. The division of the world into models, data and non-sense asks us to be honest about what we have assumed when we draw conclusions. It asks us to think hard about both our own and other people's perspectives.

*

The basis for most mathematical models for measuring skill is an equation known as the Markov assumption. Here it is:

$$P(S_{t+1}|S_t) = P(S_{t+1}|S_t, S_{t-1}, S_{t-2}, \ldots, S_1)$$

(Equation 4)

We read $P(S_{t+1}|S_t)$ in the same way as we read Equation 2 in Chapter 2. P represents the probability the world will be in state S_{t+1} and the symbol | denotes 'given'. The additional component this time is the subscript $t+1, t, t-1, \ldots$ and so on for each of the events. So, in words, $P(S_{t+1}|S_t)$ is 'the probability the world will be in state S_{t+1} at time $t+1$ given that the state of the world was previously S_t at time t'.

The key idea behind the Markov assumption is that future states of the world depend only on the recent past. Equation 4 says that the future state, at time $t + 1$, depends only on the state at the present time t, so we are assuming that the past states $S_{t-1}, S_{t-2}, \ldots, S_1$ are not relevant. To make the equation more concrete, imagine Edward, a bartender in a busy bar. Edward aims to serve his customers as quickly as possible. The number of customers can vary, but Edward strives to take as many orders as possible. In the language of maths, we let S_t denote the number of people waiting to order at minute t.

Let's put Ed to work. When his shift starts there are $S_1 = 2$ people waiting to be served. No problem at all. He pours a couple of pints for the first guy in the queue and fetches a glass of wine for the woman behind him. While he is serving these two, three more people queue up. So the number of people waiting at minute $t = 2$ is $S_2 = 3$. Ed serves them all, only to find $S_3 = 5$ more waiting at minute $t = 3$. This time he can only serve 3 of them and the 2 who remain unserved are joined by 4 more, $S_4 = 6$.

The Markov assumption says that in order to measure Ed's skill as a bartender, all we need to know is how quickly he serves his customers: we need to know how S_{t+1} depends on S_t. The number of people waiting earlier in the evening ($S_t, S_{t-1}, S_{t-2}, \ldots, S_1$) is no longer relevant for assessing his skill at this point in time. For bartending this is a reasonable assumption. Edward can serve about 2 or 3 people per minute, which is the difference between S_{t+1} and S_t.

Ed's boss, unschooled in the Markov assumption, may look out into the bar, see a lot of customers waiting and conclude that Ed isn't doing his job properly. Ed could explain the Markov assumption, and tell his boss to account for two rates, the rate at which people come into the bar and the rate at which they are being served. Ed is only responsible for the latter. Or he could just say, 'It is really busy in here tonight. Just watch how hard I'm working.' Either way, Ed is using the Markov assumption to explain the correct way to measure his skill as a bartender.

Equation 4 is different from the other equations we have seen so far, in that it doesn't give an answer straight away. In Equations 1 through 3 we put our data into the model and improved our understanding of the present or the near future. Equation 4 is an assumption. It is a step towards getting answers, but the assumption is not the answer in itself. For bartending, the Markov assumption tells us to watch the rate at which Ed serves his customers. We made a similar assumption in our football passing model: we assumed that we could forget what happened before Pogba got the ball and what happened afterwards. This assumption allowed us to measure how his specific passes helped his team.

It is important to be honest both before and after we create a model about our assumptions and whether or not they worked as we thought they would. This is the difference from the likes of Mr 'My Way', who explain their misfortune through bad luck or the mistakes of others. The skill of the modeller is to decide which events need to be included in a model and which can be safely ignored. What are the events and measurements that characterize the true state of a bar, a football team or any other type of organization?

We could very well have got our assumption wrong. While we are cheering Ed on as he mixes and serves his cocktails at top speed, his boss pokes her head out of the office for a second time. Now she sees a massive pile of dirty glasses. Ed has forgotten to put the dishwasher on! Our mistake, and Ed's embarrassing error, is due to a poor assumption. We thought the only thing that was important in the bar was the customers, and forgot about the washing up.

The manager shows Ed how to put the dishwasher on and tells him that, from now on, she will judge his skill in terms of both how quickly he cleans glasses and the rate at which he serves his customers. Together they reformulate their model so that, for example, the state $S_t = \{5,83\}$ says that the bar contains 5 waiting customers and 83 dirty glasses. Now, both Ed and the manager are happy. That is until the manager notices that Ed has forgotten to take the food that his customers have ordered out to them . . .

When considering your own life, the key to success is honesty about the aspects you are trying to improve. You might, for example, consider your salary as the most important factor for measuring your progress. The Markov assumption tells you to worry less about the pay rises you got in the past, which are no longer relevant, and more about how your current actions are improving your income. Be honest with yourself that it is salary that is important to you, but then, if your relationship starts to suffer with the long hours you are putting in, explain to your loved one the mistake you made in your assumptions. Revise your assumptions and start again.

*

The principle of verifiability that A. J. Ayer outlined in *Language, Truth and Logic* arose out of the thinking of a group of philosophers known as the Vienna Circle. At the centre of the circle was a physicist, Moritz Schlick, who chaired the group, and Rudolf Carnap, who had been a student of the great logician and mathematician Gottlob Frege.[1] The dark sulking hero of the movement was Ludwig Wittgenstein. He was a student of Bertrand Russell in Cambridge and not an active part of the circle, but it was Wittgenstein's 1922 book *Tractatus Logico-Philosophicus* that most clearly demonstrated the argument that all meaningful statements must be verified against data. Wittgenstein's seventh proposition – 'Whereof one cannot speak, thereof one must be silent' – was the ultimate 'shut up' to anyone who doubted the power of verification.

In 1933, A. J. Ayer, at the age of twenty-two, somehow managed to wangle an invitation to sit in on the Vienna Circle's discussions, and three years later his book was published. Through him, the circle's approach, which became known as logical positivism, moved from mainland Europe to England. The Second World War took Carnap and his ideas further afield to the USA. By the time the war was won, almost all the Western world had adopted the principle of experimental verification.

During the first half of the twentieth century, logical positivist thinking transformed TEN. Models had already become the focus of all scientific investigation, with Albert Einstein rewriting the laws of physics using the new mathematics. Now this approach was to be accorded a unique authority. Models and data weren't just one way of seeing the world; they were the *only* way of seeing the world.

It wasn't so much that TEN's members sat down in study circles of their own to better understand Wittgenstein, Russell, Carnap and Ayer. Some of them had read philosophy, but most of them simply followed their own reasoning about how models should be applied and came to similar conclusions as these philosophers. Remember, there is no 'TEN' in the minds of its members, so there can be no meeting to decide its principles. Logical positivism simply fit so well with the society's own thinking. It described exactly what they had been doing since de Moivre first stated the confidence equation.

TEN now entered a golden age throughout Europe. Andrey Markov (of chain fame) established the society in Russia at the turn of the century, but it was after the revolution, in the newly formed Soviet Union, that another Andrey – Kolmogorov – led its operations. Kolmogorov wrote down the axioms for probability, combining the works of de Moivre, Bayes, Laplace, Markov and others into a single unified framework. Now the code could be passed directly from teachers to small groups of students. During the summer, Kolmogorov would open up his large country villa and invite in his brightest students. Each of them was given a room on which to work on a problem. Kolmogorov toured the house, discussing each problem in turn, tuning his students' skills and developing the code. Despite the purges elsewhere, the Soviet leaders trusted TEN time and time again to drive their social ideas forward, to build their space programme and to design the new economy.

A similar spirit of intellectual freedom and trust in TEN spread across Europe. In the UK, the centre for a mathematical approach was Cambridge. It was here that Ronald Fisher rewrote the theory of natural selection in equations; that Alan Turing described his universal computing machine and laid the groundwork for computer science; that John Maynard Keynes used his undergraduate studies in mathematics to transform how governments made economic decisions; and where Bertrand Russell synthesized Western philosophy. It was also to Cambridge that David Cox came as an undergraduate student at the end of the war.

In Austria, Germany and Scandinavia, TEN tackled one physics question after another. Erwin Schrödinger wrote the equations of quantum mechanics, Niels Bohr gave the mathematics of the atom and Albert Einstein – well, Einstein did all the things that Einstein is famous for. The French, who had exiled de Moivre 200 years earlier, were not fully convinced of the principle of verifiability until after the war was over (and maybe not fully persuaded even then). And yet it was the French mathematician Henri Poincaré who laid the groundwork for the maths that would later become known as chaos theory.

TEN's model and data dichotomy took precedence over everything else: it was unswayed by religious beliefs. The Christianity which Richard Price had attributed to the judgement equation was quietly relinquished. It was unverifiable. The possibility that God might have given us mathematical truths was seen as meaningless. The idea that we might be living in the cave of Plato's allegory was nonsense. The fact that the confidence equation had its roots in gambling made no difference to its applicability and was therefore also irrelevant. All notions of religion and ethics were to be set aside and replaced by rigorous, verifiable thinking.

*

A modern group of TEN's members sit together discussing the issue of the day: the theory of relativity, or climate change, or baseball, or Brexit polls. The subjects have changed over the last one hundred years, but the nature of the discussion hasn't. It is characterized by precision. The society's members are honest about what assumptions they have made. They discuss which aspects of the world their model explains and which it doesn't. When they disagree, they compare assumptions and carefully study the data. There might be a feeling of pride on the part of the modeller who best explains the data, or a slight frustration for the modeller who is forced to admit that his model doesn't work, but they all know that this isn't about them. The greater goal is the modelling itself: to find the explanation that is least wrong.

Those who don't speak the language of models and data are either quietly admonished or politely ignored, from prejudiced politicians, screaming football coaches and angry fans to over-zealous climate activists and ignorant deniers, expounders of culture wars and Marxist fundamentalists, to Donald Trump and women-hating incels. TEN is a small group inching nearer and nearer to the truth, while the rest of humanity spirals further and further away.

*

Luke Bornn is relaxed, dressed in a T-shirt and smiling into his computer's camera. We are meeting over Skype in February 2019: he is in his brightly lit office in Sacramento, California; I am in my dark base-ment in Sweden. Behind him I can see a Sacramento Kings basketball top with his name stretched across it, while on the other side of the office is the obligatory row of academic books. The time difference between us is evidenced in our respective energy levels. As I sit trying to remember my questions, Luke rolls around his office on his chair, plucking out books from his shelf to hold up and show me on the screen.

Luke is Vice President of Strategy and Analytics at the Kings. He hasn't followed what he calls the traditional path into basketball and his current role. He points behind him and says, 'When we last announced a post for a data analyst role here, we had over a thou-sand applications, most of whose dream has been to work in sports since they were this small' – his arm now indicating the size of a four-year-old. 'My story is different,' he continues. 'I was just starting an academic position at Harvard, modelling movement of animals and climate systems, and I had a fluke meeting with Kirk Goldsberry [NBA analyst and former San Antonio Spurs strategist] where he showed me all this basketball data.'

What fascinated Luke was the richness of the observations. There was information about the players' health and fitness, data on the loading on their joints, movement patterns of all the players during training and at the games, records of the passes and the shots. Every-thing about basketball had been recorded, yet very little of it was being utilized by the team coaches.

'For me', Luke says, excitement building in his voice, 'it wasn't a "cool sports" project, it was literally the most interesting scientific challenge I had ever come across.'

He had the perfect skill-set to address that challenge and produce results quickly. In a paper presented at the MIT Sloan Sports Analyt-ics Conference in 2015 he and Kirk defined a new range of defensive metrics for basketball called 'counterpoints'. The success of Luke's approach caught the interest of the owner of A. S. Roma football club, who appointed him director of analytics. At Roma he quickly learnt how to communicate information visually, using graphs and shot plots, which proved an effective way to get mathematical ideas across to scouts and coaches. It was during his time at Roma that the club signed two world-class players in the form of forward Mohamed Salah and goalkeeper Alisson Becker, who both later left to join Liverpool and went on to win the Champions League in 2019.

'I certainly wouldn't take credit for their signings,' he tells me, 'there are too many moving parts within a football transfer. But I will say that it was after I went to the Kings that Salah and Becker moved on. I take no responsibility for selling them!'

When I tell Luke that I am interested in talking about the Markov assumption his face lights up even more than when talking about football transfers.

'We used the Markovian assumption right from the start with our defensive counterpoint paper,' he says.

Luke's counterpoint system automatically identifies who is marking whom, allowing him to measure which players come out on top in one-to-one situations. For example, during a game on Christmas Day 2013 between the San Antonio Spurs and the Houston Rockets, the defensive position of the Rockets' James Harden was best predicted by the position of Spurs' forward Kawhi Leonard. In this particular match-up it was Leonard who came out on top, scoring 20 points in the game. Luke's algorithm attributed 6.8 of those conceded points to Harden's defensive marking.

'We'd all like to know the God Model,' Luke says with a knowing smile. 'This model would tell LeBron James exactly what he should do next in order to have the best chance of scoring. But we also all know that isn't possible.'

The key to building a useful model is deciding what to take 'as given', what assumption to make. A God Model would take everything that has happened in the past as given: every training session LeBron James has attended, every game he has played, what he had for breakfast his entire life and how he tied his shoelaces before the match. This is the right-hand side of Equation 4. It takes the entire history of James's life, up to the time he takes a shot, as given. Luke's skill as a modeller lies in deciding which things can be ignored. He decides what needs to remain on the left-hand side of Equation 4 when he makes the Markovian assumption.

Luke continues. 'When we model LeBron James, we account for where he is on the court, whether he is heavily defended and where his teammates are. Then we assume that everything that came more than a few seconds before that point is irrelevant. And for the most part that assumption works.'

I ask Luke how he squares his assumption with match commentators who say things like 'He doesn't look so sharp today' or 'This player is on fire', after focusing on the last five or ten minutes of play.

'That is just bias,' Luke replies. 'The best predictor of a player making a shot is not the average over the last five shots, it is his and his opponents' positions on the court at exactly that point in time and his overall quality as a player.'

A big question in basketball is when an attacking player should pass outside of the arc to a teammate who is in a position to take a 3-point shot (shots inside the arc, nearer to the net, are worth 2 points). The Markov assumption allows Luke to replay an entire NBA season within a computer simulation. In one of these 'alternative reality' simulations, virtual players inside the arc are 'forced' to pass or dribble to a 3-point position. The results of the simulation are clear: unless a player is near the net, it is better to move the ball back out of the arc and shoot for 3 points.

It is here that James Harden shows his true value as a player. Harden has had more 50-plus scoring games than any other current player in the NBA, including LeBron James, achieving his tally in large part through 3-pointers. He has perfected a lovely movement where he dummies into the arc, looking as if he will dribble, before stepping back again and shooting for 3.

In Luke's model, James Harden's team, the Rockets, were the closest to 3-point mathematical perfection. Perhaps this isn't too surprising given that their general manager, Daryl Morey, is a graduate from Northwestern University in computer science and statistics. Another mathematician had beaten Luke to his conclusion. Harden was already putting the 3-points strategy to work, which has since been dubbed 'Moreyball'.

Basketball has become as much a battle of mathematical minds off the court as it is a sporting contest on the court. That battle is about who has the best assumptions in their model. Luke is now incorporating defensive pressure and a shot clock into his Markovian assumption, using a method called transition probability tensor, to allow him to work out – as the clock runs down – when it becomes worth throwing a desperate 2-pointer. Luke's transition probability tensor might not be as spectacular as a Harden dummy step-back-3-pointer, but it is certainly just as elegant.

*

The 'Moneyball' story of baseball coach Billy Beane, played by Brad Pitt in the movie of the same name, is one of the greatest stat-nerd adventures of our times. It tells how the general manager of baseball underdogs Oakland Athletics, with a small wage budget, assembled a team of oddball players on the basis of their statistical performance. The team went on a twenty-game winning streak.

The *Moneyball* film ends with Beane being offered, and turning down, a high-paid role at the Boston Red Sox to stay at his beloved A's. It is a romantic ending to the film, but it is not particularly reflective of what happened in baseball after Oakland's success.

Beane is an ex-player, not a statistician or economist by training. Most often, though, when the owners of other baseball clubs looked to replicate Beane's success, it wasn't to open-minded ex-pros like him that they turned; it was directly to the mathematicians themselves. Bill James, the creator of the statistical approach applied by Beane, did accept a role at the Boston Red Sox, and has been working for them since 2003. The Red Sox also appointed a maths graduate, Tom Tippett, as their director of baseball information services.

Another team that built success around statistics are the Tampa Bay Rays, who employed Doug Fearing, a Harvard Business School assistant professor in Operational Research in 2010. During the five years Doug worked there, the Rays qualified for the Division Series three times, with one of the lowest wage bills in Major League Baseball.[2] Doug went on to the LA Dodgers, where he had twenty people working in his analysis group, at least seven of whom had Masters or PhDs in statistics or maths. They analysed everything from defensive positioning and the batting order to the length of players' contracts.

I caught up with Doug shortly after he had left the LA Dodgers in February 2019 to start his own sports analytics company. The first thing I asked him was if he was a big baseball fan.

'I would say, relative to some of the people who work in sports, I am maybe not a fan,' Doug joked, 'but relative to the population, I would say "absolutely", I'm a big fan.' Doug had followed the Dodgers throughout his life and working for them was his dream job.

Modern baseball analytics has its roots in the work of amateur statisticians with an interest in the sport. When I asked Doug about the Moneyball theory, he told me that 'a lot of the Oakland A's success was down to Paul DePodesta [represented by Jonah Hill's character in the *Moneyball* film] taking methods that were already used in the public sphere and implementing them in internal decision-making within the club'.

Doug described how general managers in baseball with successful playing careers and an intuition for the game had been gradually replaced by university graduates with an Ivy League background and an understanding of data.

'Baseball can be approximated as a series of one-on-one matches between the batter and the pitcher,' Doug told me, 'so the Markov assumption works very well in so many situations.' The ease with which the Markov assumption can be applied makes baseball easier to analyse than other sports. As a result, the mathematicians have taken over more quickly.

Doug talked enthusiastically about the classic early scientific articles on baseball analytics from the 1960s and 1970s. In his article published in 1963, George R. Lindsey used a statistical model to answer questions, such as when a runner should try to steal base and when the fielding team should place their infield close to the batter. His Markovian assumption was that the number of men out and the arrangement of runners between bases was the state of the game. He found optimal batting and fielding strategies by testing his model against data, manually collated by his father Colonel Charles Lindsey, of 6,399 half-innings from the 1959 and 1960 seasons. Lindsey prefaced the conclusions of his article, 'It must be reiterated that these calculations pertain to the mythical situation in which all players are "average".'[3]

This slightly overstated honesty, to see his own model simultaneously as a myth and as something useful, is the mark of a true mathematical modeller. Accurate reporting of assumptions is just as important as accurate reporting of results.

These mathematical models were, on the whole, discovered by those working outside the game: men and women who were fascinated by the statistics and wanted to explain them. Once the power of numbers has been acknowledged in a sport, those who know the code are welcomed in and those without the skills are asked to move on. In baseball the transition is complete. In basketball it is underway and in football it is beginning to take hold. Liverpool FC, who Luke credited with signing the best players at Roma during his time there, is owned by the American businessman John W. Henry, the man who brought Bill James to the Boston Red Sox. Liverpool hired theoretical physicist Ian Graham to help with their recruitment. When they won the Champions League in 2019, the *New York Times* interviewed him and his fellow analyst and physicist William Spearman about their roles.[4] By sanctioning the interview, the club were happy to give them at least part of the credit for the team's progress. The 2018–2019 Premier League title-winners Manchester City also have a large team of data analysts and, as we already know, so too do the 2019 La Liga champions, Barcelona. Other teams, most notably Manchester United, don't seem to have caught on yet. It appears that they have no idea how much Paul Pogba is really worth to them.

The red side of Manchester has been warned. The same rules that apply elsewhere in life, apply in sport too. Models win. Nonsense loses.

*

When you set out to measure your own skill, or the skill of others, you need to make your assumptions clear. What is the state of affairs before you act and what is their state following your actions? Decide on specific areas in your life you would like to improve. Maybe you would like to learn more maths or to go out running more often? Be honest about where you are now: which equations you do and don't know or how many kilometres you run each week. That is the current state of affairs. Write that down and start working on your performance. One month later, look again at where you are. The skill equation tells you to be honest about what you assumed before you started. Don't justify your failure by claiming you were trying to achieve something else and don't downplay your successes by becoming distracted by failures elsewhere in your life. But do re-evaluate your assumptions before you continue. Re-evaluate how it is that you really want to improve. Don't dwell on history. Use the Markov assumption to forget the past and concentrate on the future.

Speaking to Luke Bornn makes me aware that I have a few skills that I need to improve. I need to be better and more patient when talking to Mr 'My Way'. When we set aside my cartoon representation of him at the start of the chapter, Mr 'My Way' does have something to offer. He has experience and drive. He is good with people. And he knows and loves the sport he is involved in. How can I stop Mr 'My Way' from talking so much nonsense and help him focus on models and data instead?

Luke told me that it is really common for him to sit in scouting meetings where the discussion will start with one scout saying, 'Do you like this guy?', referring to a player they watched recently.

'Yeah, I like him,' says another scout.

'Yeah, I love this guy. He's a baller,' says a third.

'Well, I don't like him,' says the first scout.

In these situations, Luke tries to use statistics to give a bit of context. He says to the third scout, who was convinced his man was a baller, 'Well, you saw this guy November 22nd and the stats show he had the game of his life. So . . .'

From there the discussion can take a new, more informed, direction. Maybe they can watch a video of the player together and discuss how representative it is of his play.

One thing that impressed Luke going into sports was how open the staff at sports clubs are. Every scout he encounters wants every piece of available information. Just like mathematicians, they are hungry for data. Luke tries to provide a way of organizing this information in the form of an underlying model. 'We try to be honest about what our models say,' Luke told me. 'We lay everything out on the table and tell the scouts our assumptions. This becomes the foundation for their discussions.'

He provides the rest of his organization with statistical summaries, charts, newspaper reports, anything and everything they need. He tries not to use the word 'data' when talking, which tends to be used in a debate about the reliance on data or on human knowledge; instead he sees himself as a provider of information. As Luke asked me rhetorically, 'Who doesn't want more information?'

I was intrigued by the way he saw himself primarily as a resource, and couldn't help reflecting that, 'the way you talk, you put yourself on a lower level than the scouts . . .'

'Maybe, in a way,' replied Luke after some thought. 'I don't need to be the one smart person at the Kings. I would much rather be someone who makes everyone else smarter.'

This modesty is, in my experience, a characteristic of many of the best applied mathematicians and statisticians.

I thought back to my discussion with Sir David Cox. When we had talked about the concept of genius, he had become very thoughtful. '"Genius" isn't a word I use,' he told me, 'it's a very strong word.' He thought some more and continued, 'I've never heard anyone use the word "genius" except maybe for R. A. Fisher', a reference to the Cambridge statistician, widely recognized as the father of modern statistics. 'And even then,' he added, 'they probably used it slightly sardonically. Maybe this is very English, but I'd regard the word as over the top.' Sir David went on to name a few people he might consider as geniuses: Picasso, Mozart or Beethoven.

The term 'genius' is often employed when talking about the applications of mathematics: the genius of Albert Einstein in physics, of John Nash in economics and of Alan Turing in computer science. Their contributions are certainly hugely impressive, but the term doesn't properly reflect the way their work should be seen. It tends to make it inaccessible to the rest of us and turns the mathematician himself into a Mr 'My Way', who sees himself as smarter than you or me.

There are geniuses at Barcelona. They are Lionel Messi, Sergio Busquets, Samuel Umtiti and the others. They see things we'll never see. Their performances create an art that few others can replicate.

The members of TEN are not geniuses. We produce ideas that are repeatable and measurable. We sort and organize data. We clear away the non-sense. We provide models. And, when we are at our best, we do it invisibly.

5

The Influencer Equation

$$A \cdot \rho_\infty = \rho_\infty$$

Have you ever thought about the probability that you happen to be you, and not someone else? I don't just mean someone slightly different – a person who has or hasn't been to Disneyland or has or hasn't seen all the *Star Wars* films – I mean someone completely different: someone born in another country, perhaps, or even in another time.

The population of our planet is nearly 8 billion. This means that the probability that you are the particular person that you happen to be is 1 in 8 billion. The probability of getting all 6 numbers right in a 49-ball lottery, like the one run in the UK, is around 1 in 14 million. The probability of winning the lottery with a single ticket is 570 times *larger* than the probability of you being you.

I sometimes imagine a universe in which I wake up as a random person every day. The calculation above tells us that we can pretty much forget the possibility that we wake up as ourselves two days in a row – the chances of that would also be 1 in 8 billion, but what is the probability that we wake up in the same city as we went to sleep? The city Uppsala in Sweden where I live has around 200,000 inhabitants. On a global scale this means that the probability that I wake up here tomorrow is only 1 in 40,000. If I continued my journey, waking up as a random person every morning for the next fifty years, the probability I would land in Uppsala at some point is about 50%. A coin flip would determine whether I ever see sunrise in my home town again.

On my random journey I would spend a day in London and a day in Los Angeles once every two years. I'd visit New York, Cairo and

Mumbai almost one day per year.[1] Greater Tokyo, with its population of 38 million, would be my new home almost twice per year. While the probability of waking up in any specific city remains small, I would be more likely to wake up in a densely populated urban area than in the countryside. The most likely place for me to wake up each day is China, closely followed by India. If there is some stability to be found in all the noise of random body inhabitation it is to be found in these two countries. With combined populations of 2.8 billion, I would spend about two and a half days of the week living in one of these countries. Africa would be my home once a week on average, compared to visits to the USA of just over once a month. My journey, which in all likelihood would never take me back to my starting point, reminds me that I am both uniquely improbable and impossibly unimportant.

Now imagine that instead of waking up as a person selected from the world's population, I wake up as one of the people I follow on Instagram. I'm not a big user of the picture-sharing social network, and only follow the few friends who have taken the time to look me up, so I will wake up as one of them: maybe a friend from school or an academic colleague at a different university. I will take over their body for a day, learn how it is to be them, maybe even send my old self a message, before spending a night in their bed and waking up as another person, randomly selected from the people they follow.

I might even wake up as David Sumpter again. Typical Instagram users tend to follow between 100 and 300 people, so given I have a mutual (follow/follow back) relationship with all the people I follow, there is a reasonable chance (around 1 in 200) that I will end up spending another day as myself. Whether I become myself again or not, I will very likely spend a few days travelling through my social group, friends of friends and people who are like me in culture and background.

Then something happens that changes my life for ever. I wake up as Cristiano Ronaldo. Well, perhaps not exactly Ronaldo. Maybe it will be Kylie Jenner, or Dwayne 'The Rock' Johnson, or possibly Ariana Grande. While the particular celebrity in question might vary, the transformation to stardom itself is guaranteed. Within a week or so of starting my journey, I will have become one of the best-known people on Instagram. These celebrities, with hundreds of millions of followers, are followed by those people in my immediate social network and sooner rather than later I will jump into their bodies.

I could very well stay in this celebrity world for a week or even longer. Cristiano follows Drake, Novak Djokovic, Snoop Dogg and Steph Curry, moving me backwards and forwards between sports stars and rappers. From Drake, I hop into Pharrell Williams and then Miley Cyrus. She takes me on to Willow Smith and Zendaya. Now I am moving freely through a world of musical artists and film stars.

Then, after a fortnight of fame, another transformation occurs, one that's even more dramatic than waking up as Snoop Dogg. One morning, after spending a day of action-movie filming, I wake up as a school friend of Dwayne 'The Rock' Johnson. At this point, I realize a terrible truth. I am lost. There is almost no chance at all that I will ever be David Sumpter again. Quite soon, I will hop back into the celebrity circle again, hanging with the stars and sharing pictures of my semi-naked body. These periods will be interspersed with journeys to B-list celebrity, and some brief times in the bodies of normal people again before returning to the shiny world of stardom.

The probability that I will become myself again tomorrow is very small indeed – maybe one in a trillion, or possibly even less. All random Instagram journeys point towards celebrity and stay there.

*

The single most important equation of the twenty-first century is as follows:

$$A \cdot \rho_\infty = \rho_\infty$$

(Equation 5)

Forget the billion-dollar earnings of logistic regression in gambling, this equation is the basis for a trillion-dollar industry. It is Google. It is Amazon. It is Facebook. It is Instagram. It is at the heart of every Internet business. It is the creator of superstars and the suppressor of the everyday and the menial. It is the maker of influencers and the anointer of social media queens and kings. It is the cause of our ceaseless need for attention, our obsession with our self-image, our frustration and fascination with fashion and the drivers of celebrity. It is why we feel lost in a vast sea of advertising and product placement. It has shaped every part of our online lives.

It is the influencer equation.

You might imagine that such an important equation is difficult to explain or to understand. It isn't. In fact, I explained it just now when I imagined I had become Ronaldo, The Rock or Willow Smith. All we need to do is relate the symbols A (called a connectivity matrix) and ρ_t (a vector giving the probability of being each of the people in a social network at time t) to the tour we just went on of the world's population.

To visualize connectivity matrices, imagine a spreadsheet where both the rows and columns are names of people. The cells in this spreadsheet give the probability of waking up tomorrow as a different person. Imagine a world with only five people: myself, The Rock, Selena Gomez and two people I've never heard of, who I'll call Wang Fang and Li Wei. If I assume, as I did in my first thought experiment, that I'll wake up as a random person each day, then the matrix A will look like this:

$$A = \begin{array}{c} \begin{array}{ccccc} DS & TR & SG & WF & LW \end{array} \\ \begin{pmatrix} 1/5 & 1/5 & 1/5 & 1/5 & 1/5 \\ 1/5 & 1/5 & 1/5 & 1/5 & 1/5 \\ 1/5 & 1/5 & 1/5 & 1/5 & 1/5 \\ 1/5 & 1/5 & 1/5 & 1/5 & 1/5 \\ 1/5 & 1/5 & 1/5 & 1/5 & 1/5 \end{pmatrix} \begin{array}{c} DS \\ TR \\ SG \\ WF \\ LW \end{array} \end{array}$$

The labels on the rows and columns of the matrix are the initials of the world's five inhabitants. Each day, I look up the column of the person whose body I am currently in and then the values in each of the rows tells me the probability I will be that particular person tomorrow. The fact that all the entries are 1/5 reflects the fact that tomorrow I can become any one of the five people (including myself again).

If, on the other hand, I assume, as I did in my second thought experiment, that I wake up as someone I've followed on Instagram, then the matrix A will take a different form. To make this a bit interesting, let's assume that The Rock got stuck on a maths problem and decided to follow me on Instagram. Moreover, we'll assume that Selena Gomez met Fang and Wei at one of her concerts, thought they looked cute together (I forgot to mention that Fang and Wei are a couple) and followed them. Everyone follows Selena and The Rock, of course. Now we have:

$$A = \begin{array}{ccccc} \text{DS} & \text{TR} & \text{SG} & \text{WF} & \text{LW} \\ \begin{pmatrix} 0 & 1/2 & 0 & 0 & 0 \\ 1/2 & 0 & 1/3 & 1/3 & 1/3 \\ 1/2 & 1/2 & 0 & 1/3 & 1/3 \\ 0 & 0 & 1/3 & 0 & 1/3 \\ 0 & 0 & 1/3 & 1/3 & 0 \end{pmatrix} & \begin{array}{l} \text{DS} \\ \text{TR} \\ \text{SG} \\ \text{WF} \\ \text{LW} \end{array} \end{array}$$

When I am David Sumpter, there are only two people I can be tomorrow: Selena or The Rock, so I have 1/2 in each entry of my column. The Rock is the same, while the others each have three people they can transform into. The diagonal line of zeros in the matrix means we can't be the same person two days running, because we don't follow ourselves.

Notice that I have used the Markov assumption (Equation 4 from Chapter 4), to create my model. I have assumed that the person I was two days ago has no influence on who I am tomorrow. In fact, matrix A can also be referred to as a Markov chain, because A tells us how the next step in a chain of events will depend solely on the current event.

We now start marking steps through the days by following the A chain. Assuming, on the first morning, I wake up as David Sumpter, then we can calculate who I will be tomorrow as follows:

$$\begin{array}{ccccc} \text{DS} & \text{TR} & \text{SG} & \text{WF} & \text{LW} \end{array} \quad \text{Day 1} \quad \text{Day 2}$$

$$\begin{pmatrix} 0 & 1/2 & 0 & 0 & 0 \\ 1/2 & 0 & 1/3 & 1/3 & 1/3 \\ 1/2 & 1/2 & 0 & 1/3 & 1/3 \\ 0 & 0 & 1/3 & 0 & 1/3 \\ 0 & 0 & 1/3 & 1/3 & 0 \end{pmatrix} \cdot \begin{pmatrix} 1 \\ 0 \\ 0 \\ 0 \\ 0 \end{pmatrix} = \begin{pmatrix} 0 \\ 1/2 \\ 1/2 \\ 0 \\ 0 \end{pmatrix} \begin{array}{l} \text{DS} \\ \text{TR} \\ \text{SG} \\ \text{WF} \\ \text{LW} \end{array}$$

I explain how we multiply matrices in detail in this endnote,[2] but the most important thing to notice are the two columns of numbers in brackets either side of the equals sign. These are called vectors, and each row of the vector contains a single number between 0 and 1 that measures the probability that I am a particular person on a particular day. On day 1, I am David Sumpter, so my row is 1 and everyone else is zero. On day 2, I am either Selena Gomez or The Rock (one of the two people David Sumpter follows), so the vector contains 1/2 for them and zeros for everyone else.

Things start to get more interesting on day 3. Now we have:

$$
\begin{array}{ccccc}
\text{DS} & \text{TR} & \text{SG} & \text{WF} & \text{LW}
\end{array}
\quad \text{Day 2} \quad \text{Day 3}
$$

$$
\begin{pmatrix}
0 & 1/2 & 0 & 0 & 0 \\
1/2 & 0 & 1/3 & 1/3 & 1/3 \\
1/2 & 1/2 & 0 & 1/3 & 1/3 \\
0 & 0 & 1/3 & 0 & 1/3 \\
0 & 0 & 1/3 & 1/3 & 0
\end{pmatrix}
\cdot
\begin{pmatrix}
0 \\ 1/2 \\ 1/2 \\ 0 \\ 0
\end{pmatrix}
=
\begin{pmatrix}
1/4 \\ 1/6 \\ 1/4 \\ 1/6 \\ 1/6
\end{pmatrix}
\begin{matrix}
\text{DS} \\ \text{TR} \\ \text{SG} \\ \text{WF} \\ \text{LW}
\end{matrix}
$$

I can be any one on our planet of five people. I am more likely to be David Sumpter or Selena Gomez, but I could also, with a probability of 1/6 each, be The Rock or one of Selena's Chinese fans. Let's multiply again to find out who I'm likely to be on day 4.

$$
\begin{array}{ccccc}
\text{DS} & \text{TR} & \text{SG} & \text{WF} & \text{LW}
\end{array}
\quad \text{Day 3} \quad \text{Day 4}
$$

$$
\begin{pmatrix}
0 & 1/2 & 0 & 0 & 0 \\
1/2 & 0 & 1/3 & 1/3 & 1/3 \\
1/2 & 1/2 & 0 & 1/3 & 1/3 \\
0 & 0 & 1/3 & 0 & 1/3 \\
0 & 0 & 1/3 & 1/3 & 0
\end{pmatrix}
\cdot
\begin{pmatrix}
1/4 \\ 1/6 \\ 1/4 \\ 1/6 \\ 1/6
\end{pmatrix}
=
\begin{pmatrix}
6/72 \\ 23/72 \\ 23/72 \\ 10/72 \\ 10/72
\end{pmatrix}
\begin{matrix}
\text{DS} \\ \text{TR} \\ \text{SG} \\ \text{WF} \\ \text{LW}
\end{matrix}
$$

Now we start to see that celebrities take a more central role. Reading the values from the vector for Day 4, we see that the probability I am either The Rock or Selena Gomez is 23/72, almost four times the probability I am David Sumpter, which is 6/72.

Every time we multiply by the connectivity matrix A we move one day into the future. The question that motivates my travels through the world's population is, over a longer period of time how often will I spend being each of these five people?

This is the question that Equation 5 answers. To see how, we replace the matrices and vectors containing numbers with symbols. We already saw on a previous page that the matrix is called A. Let's now call the vectors ρ_t and ρ_{t+1}, allowing us to rewrite the matrix above in a much more succinct form:

$$A \cdot \rho_t = \rho_{t+1}$$

ρ_t is a vector giving the probability of being each of the people in a social network at time t. We use the subscript t to denote time in the same way as we did in the last chapter. Up to now we have seen that:

$$\rho_1 = \begin{pmatrix} 1 \\ 0 \\ 0 \\ 0 \\ 0 \end{pmatrix}, \rho_2 = \begin{pmatrix} 0 \\ 1/2 \\ 1/2 \\ 0 \\ 0 \end{pmatrix}, \rho_3 = \begin{pmatrix} 1/4 \\ 1/6 \\ 1/4 \\ 1/6 \\ 1/6 \end{pmatrix} \text{ and } \rho_4 = \begin{pmatrix} 6/72 \\ 23/72 \\ 23/72 \\ 10/72 \\ 10/72 \end{pmatrix} \begin{matrix} \text{DS} \\ \text{TR} \\ \text{SG} \\ \text{WF} \\ \text{LW} \end{matrix}$$

Now we come to Equation 5, which I'll repeat here because it is a while since we've seen it:

$$A \cdot \rho_\infty = \rho_\infty$$

We have replaced both the t and the $t+1$ in the equation above with the infinity sign, ∞. By doing this, we are saying that as we go infinitely forward in time there is no difference between t and $t+1$. Think about this for a second. The implication here is that, provided we have hopped between bodies for enough days, then it doesn't matter if we hop one more hop: the probability of being in a particular body is the same and indicated by ρ_∞. We call ρ_∞ the stationary distribution because it indicates how our time in each state, in the body of each person, as time moves forward and who we were initially, is forgotten.

Equation 5 gives the probability that I will wake up as a particular person on some day in the distant future. Now the only thing left to do is solve the equation. For the five-person universe in which I currently reside, we find that:

$$
\begin{pmatrix}
0 & 1/2 & 0 & 0 & 0 \\
1/2 & 0 & 1/3 & 1/3 & 1/3 \\
1/2 & 1/2 & 0 & 1/3 & 1/3 \\
0 & 0 & 1/3 & 0 & 1/3 \\
0 & 0 & 1/3 & 1/3 & 0
\end{pmatrix}
\cdot
\begin{pmatrix}
8/60 \\
16/60 \\
18/60 \\
9/60 \\
9/60
\end{pmatrix}
=
\begin{pmatrix}
8/60 \\
16/60 \\
18/60 \\
9/60 \\
9/60
\end{pmatrix}
\quad
\begin{array}{l}
\text{David Sumpter} \\
\text{The Rock} \\
\text{Selena Gomez} \\
\text{Wang Fang} \\
\text{Li Wei}
\end{array}
$$

Notice that the two vectors on the left and right of the equals sign are the same and that their elements add up to one. This means that however many times I multiply them by the transition matrix A I will get the same result. These are, in the long term, the probability that I will be each person.

The conclusion? I am twice as likely to wake up as The Rock as David Sumpter and I am even more likely to wake up as Selina Gomez. I am also more likely to wake up as Wang Fang and Li Wei than as David, but only very slightly so. We can turn these probabilities upside down to see how long I will spend in each person's shoes. Sixty days is about two months, and the stationary distribution tells us that I will spend 8 of those days as David, 16 as The Rock, 18 as Selena, and 9 days each as Fang and Wei. More than half of my life will be spent as a celebrity as my days move off towards infinity.

*

Obviously, we don't wake up in other people's beds every morning, but Instagram does give us a glimpse into each other's lives. Each photo looked at is a moment with the person you follow, a few seconds experiencing how it is to be someone else.

Twitter, Facebook and Snapchat also allow us to spread information and influence how our followers feel and think. The stationary distribution, ρ_∞, measures this influence, not only in terms of who follows whom, but also in terms of how quickly an idea or a meme spreads between users. Those people with a large value in the vector ρ_∞ are more influential and spread memes faster. Those of us with a low value of ρ_∞ are less influential.

This is why Equation 5, the influencer equation, has been so valuable to Internet giants. It tells them who are the most important people on their network, without the company knowing anything about who these people actually are or what they do. Measuring the influence of people is just a question of matrix algebra, which a computer performs unthinkingly and uncritically.

The original online use of the influencer equation was by Google in their PageRank algorithm, just before the turn of the century. They calculated the stationary distributions for websites, assuming that users clicked randomly on links on the sites they visited in order to choose the next site they visited. Those sites with a high value of ρ_∞ were placed higher in their search results. Around the same time, Amazon started to build a connectivity matrix, A, for its business. Books, and later toys, films, electronics and other products that were bought together were connected inside the matrix. By identifying strong connections in their matrix, Amazon could make recommendations for its users, the so-called 'Also Liked' suggestions. Twitter uses the stationary distribution of its network to find and suggest who you should follow. Facebook has used the same ideas in sharing news and YouTube has used it to recommend videos. Over time, the approach has evolved and extra details have been added, but the basic underlying tool for finding the influencers on social media remains the connectivity matrix A and its stationary distribution ρ_∞.

During the last two decades this has led to an unexpected consequence. A system that was initially set up to measure influence has evolved to become the *creator* of influence. Algorithms based on the influencer equation decide which posts should feature prominently on social media feeds. The idea is that if someone is popular, then more people want to hear from that person. The result is a never-ending feedback loop. The more influential a person is, the more prominence the algorithm gives them and their influence then grows further.

A former Instagram employee told me that the company's founders were initially very reluctant to use algorithms and mathematics within their business. 'They saw Instagram as very niche, very artistic, and they viewed algorithms as inauthentic,' he told me. The platform was for the sharing of photographs between close friends. That all changed when Facebook took over. 'Over the last couple of years, the platform has become drastically different. One per cent of its users now have over 90% of its followers,' my contact told me.

Instead of encouraging users to follow only their friends, the company solved the influencer equation for its social network. It promoted the most popular accounts. The feedback began and the celebrity accounts grew and grew. So too did the company, to over 1 billion users. Just like every other social media platform before it, once Instagram was using the influencer equation, its popularity exploded.

*

The mathematics used to build our social networks had already been created long before the applications became possible. The influencer equation was not invented by Google; its origins can be traced back to Markov, who created his assumption as a way of looking at chains of states, where each new state depended only on the last, as it does on my random journey through Instagram.

When I solved Equation 5 for my own five-person online world, I was a bit lazy. I found the answer – the length of time I would spend as each person in the long term – by repeatedly multiplying the vector ρ_t by the matrix A until it no longer changed. In doing so I found ρ_∞. This method gets the correct answer – eventually – but is not very elegant. Nor is it the method used by Google. Over one hundred years ago, two mathematicians, Oskar Perron and Georg Frobenius, showed that, for every Markov chain A, there is guaranteed to be a unique stationary distribution ρ_∞. So, irrespective of the structure of a social network, we can always find out how much time we will spend as each person if we walk randomly between people. This stationary distribution can then be found using a technique called Gaussian elimination, a method that, like the Normal curve, is credited to Carl Friedrich Gauss but has its roots elsewhere. Chinese mathematicians have been doing Gaussian elimination for over two millennia. The method involves pivoting and rearranging the rows of the matrix A in order to find a solution for ρ_∞, making the influencer calculation fast and efficient, even for networks of millions of individuals.

Throughout the twentieth century TEN collected results about the properties of networks in a field of research known as graph theory. As early as 1922, Udny Yule described the mathematics behind the growth of popularity on Instagram in terms of a process later called 'preferential attachment': the more followers a person has, the more they attract and celebrity grows. Then, at the start of the twenty-first century, preceding the arrival of Facebook by only a few years, this same area of research exploded and became known as network science: it now described the spread of memes and fake news, the way social media creates a small world, in which everyone is connected by less than six degrees of separation, and the potential for polarization.[3]

TEN was ready. Its members were among the founders and the first employees of the future social media giants. The influencer equation lay at the centre of their businesses. The salaries offered to those with the skills were enough to lure even the most idealistic of the society's members. And, more importantly, the jobs gave them the freedom to think creatively, to imagine new models and to put them into practice.

TEN's members were soon tasked with finding out how we responded to our social network. They manipulated Facebook feeds to see how users reacted to receiving negative news only; they created social media campaigns to encourage people to vote in elections; and they constructed a filter so that users saw more of the news they were interested in. They controlled what we saw, deciding whether we saw posts from friends, news (fake and real), celebrities or adverts. It was TEN that became the influencers, not through what they said, but in the decisions they made about the ways we should connect to each other. They even understood things about us that we didn't even know ourselves . . .

*

Your friends are probably more popular than you are. I don't know anything about you as a person, and I don't want to judge you unfairly, but I can say this with some certainty.

The mathematical theorem known as the friendship paradox states that the majority of people on every social network, including Facebook, Twitter and Instagram, are less popular than their friends are.[4] Let's start with an example. Imagine we remove The Rock from the social network we looked at earlier. Now we have four people – me, Selena Gomez, Fang and Wei – with 0, 3, 2 and 2 followers respectively. Fang and Wei are probably feeling quite popular, hanging in a clique with Gomez, but I've got a surprise for them. I ask them to count the average number of followers of the friends of each person in our network. I only follow Selena Gomez and she has 3 followers, so the average number of followers of my friends is 3. Gomez follows 2 people, both of whom have 2 followers, so the average follower for her friends is 2. Fang and Wei follow each other and Gomez, so on average their friends have 2.5 followers. Thus, the average followers of friends on the network is $(3 + 2 + 2.5 + 2.5)/4 = 2.5$. Only Selena Gomez has more friends than her followers. Fang and Wei are, like me, below average.

The reason for the friendship paradox lies in the difference between picking a person at random and picking a friendship relationship at random (see Figure 5). Let's start by picking one person totally at random. Their expected, or average, number of followers is the sum of the number of followers of each person on the network divided by the total number of people who use the platform. On Facebook this is about 200. For the Gomez network this is $(0 + 3 + 2 + 2)/4 = 1.75$. We call this the average in-degree of the nodes (the people) in the graph (the social network). For the Gomez network we have already shown that the average followers of friends is 2.5, which is greater than the 1.75 we got when we just looked at the average number of followers.

The same result applies if we add The Rock back into the network. It even applies if Selena Gomez follows me too. In fact, the friendship paradox can be proved to be true for any network in which everyone follows exactly the same number of people. The proof is as follows. First, pick one person at random from the whole network; then pick someone that they follow. Thought about in another way, by picking two linked people, what we are actually doing is picking a random link from all the followership relationships in the network. In graph

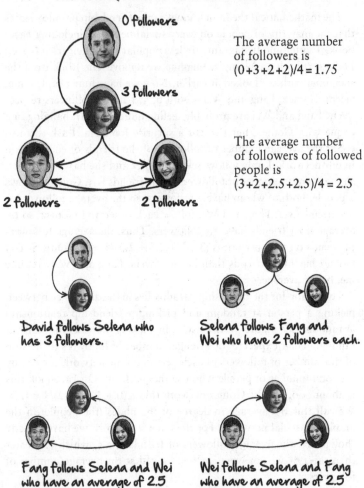

O followers

The average number
of followers is
$(0+3+2+2)/4 = 1.75$

3 followers

The average number
of followers of followed
people is
$(3+2+2.5+2.5)/4 = 2.5$

2 followers   2 followers

David follows Selena who
has 3 followers.

Selena follows Fang and
Wei who have 2 followers each.

Fang follows Selena and Wei
who have an average of 2.5
followers.

Wei follows Selena and Fang
who have an average of 2.5
followers.

Figure 5: The friendship paradox for four people

theory, these links are called the edges of the graph. Now, since popular people have (by definition) more edges going into them, we are more likely to find a popular person on the end of any given edge than if we had simply chosen a person at random. Thus, a randomly chosen friend of a randomly chosen person (the person at the end of the edge) is likely to have more friends than a randomly chosen person, and the friendship paradox holds.[5]

That is the mathematical theory. So how does it work out in practice? Kristina Lerman, Research Associate Professor at the University of Southern California, decided to find out. She and her colleagues took the network of Twitter users in 2009 (at an early stage in the life of the social network when it had only 5.8 million users) and looked at follower relationships.[6] They found that the people followed by a typical Twitter user had around ten times as many followers as they themselves had. Only 2% of users were more popular than their followers.

Lerman and her colleagues went on to find another result that completely defies our intuition. She found that the followers of a randomly chosen Twitter user were on average twenty times better connected than they are! While it appears reasonable that the people we follow are popular – many of them are celebrities after all – it is much more difficult to grasp how the people that follow us have become more popular than us. If they are following *you*, how can they be more popular? It doesn't seem fair.

The answer lies in our tendency to create mutual followed/follower relationships. There is a social pressure to become 'mutuals' when someone follows you. It is rude not to follow them back. On average, the people who follow you on Instagram, or send you friend requests on Facebook, are also likely to have sent similar requests to others. The result is that these people make up a larger part of our social network. It gets worse. The researchers also found that your friends post more, get more likes, more shares, and they reach more people than you do.

Once you have accepted the mathematical unavoidability of not being popular, your relationship with social media should start to improve. You are not alone. Kristina Lerman and her colleagues' study estimated that 99% of Twitter users are in the same situation as you are. Indeed, popular people may well be even worse off than you are. Think about it. In their never-ending search for a better social position, the 'cool kids' are trying to become mutuals with people who are more successful than they are. The more they do so, the more they end up surrounded by those who are more popular than they are. It's a small consolation, but it is good to know that even those who appear successful probably feel the same way you do. With the possible exception of Piers Morgan and J. K. Rowling, the remaining 1% of Twitter users are either PR-managed celebrity accounts or, very likely, people who have been driven half-crazy by the need to keep up online appearances.

*

I am not going to be one of those people who tell you to switch off your social media. The mathematician's mantra is not to give up. Instead, it is to break things down into their three component parts: data, model and nonsense.

I recommend that you start today. The first thing to do is look at the data. Check for yourself how many followers or mutuals your friends have on Facebook or Instagram. I just did it on Facebook and 64% of my friends are more popular than I am. Then remember the model. Popularity online is generated by feedback whereby already popular people seek out and get more followers. It is a statistical illusion generated by the friendship paradox. Then cut out the nonsense. Don't feel sorry for yourself or get jealous of others: just realize that we are all part of a network that distorts our self-worth in all sorts of different ways.

Psychologists write and talk about our cognitive biases, in which individuals or society experiences a subjective reality that deviates from how the world actually is. The list of these biases is growing longer and longer: the hot-hand fallacy, the bandwagon effect, survivorship bias, confirmation bias, the framing effect. Members of TEN certainly don't deny the existence of these biases, but, for them, the limitations of human psychology aren't the most important point. The question is how to remove the filter and see the world more clearly. To do this, they imagine 'what if' scenarios. What if I woke up as a random person each day? What if I travelled through Snapchat like an Internet meme? What if I only read news offered to me by Facebook or movies recommended by Netflix? How would the world look to me then? And how does that differ from a 'fairer' world where I pay equal attention to all the information available to me?

TEN asks you to imagine incredible, fanciful scenarios. These 'what if' scenarios then become mathematical models. From there the cycle can begin. Model is compared to data and data is used to refine the model. Slowly but surely the members of TEN can remove the filter and reveal our social reality.

*

Lina and Michaela open up their Instagram accounts and show their phones to me. 'Is that an advert or a selfie?' I ask Lina.

Lina is showing me a picture of a local baker, holding up a tray full of heart-shaped cakes for the camera. It feels genuine, but the idea is to encourage their followers to come past the shop. She replies that it is a selfie, but she has classified the account as a company.

Lina and Michaela are working on their undergraduate project in maths.[7] They are studying how Instagram presents the world to them. Just before they started their project, Instagram updated the algorithm which decides the order pictures are presented in – again. The company claimed that the focus had shifted towards prioritizing photos from friends and family.

As a result, many influencers felt threatened. Swedish Instagramer and social media guru Anitha Clemence (65,000 followers) said, 'It's psychologically stressful seeing my followers disappear. I'm nearly 40, what must it be like for all the young influencers?'[8]

Clemence felt that she was working hard 'for [her] followers' and the new algorithm wasn't getting the message out to them. To test the limits, she posted a picture of herself with her new partner which could easily be misinterpreted as showing that she was pregnant. This picture was spread widely on the platform before Clemence revealed that its primary aim was to test which pictures worked online and which didn't. Claiming you are pregnant on Instagram seems to work if you want to get more engagement.

Although the reaction to one picture of a fake pregnancy tells us very little, Clemence can be said to be conducting an experiment of sorts. Kelley Cotter at Michigan State University found that many

Instagram influencers are trying to understand and manipulate the algorithm.[9] They openly discuss the costs and benefits of liking and commenting on as many posts as possible, or talk about the best times to post, conducting A/B tests of different strategies (remember the betting equation). Often these influencers want to determine whether the site is shadowing them by placing them further down their followers' feeds. When Instagram changed their algorithms, many of these influencers took to social media to declare #RIPInstagram.

Lina and Michaela now planned to look more thoroughly at the Instagram algorithm, from their own position as typical users. For the next month, they will open up their accounts just once a day, at 10 a.m., and they will look at the order the pictures are presented to them and note the type of each post and the poster. In that way they can test the influencers' hypothesis that they have been shadowed or deprioritized by the algorithm.

'I think opening less frequently will be good for us,' says Michaela, referring to the fact that they had to collect data (look at their posts) just once a day. Like many of us, these two young women check their social media more than they really want to.

The challenge is to reverse-engineer the Instagram algorithm: to find out what (if anything) Instagram is hiding from them. In mathematics, this is called the inverse problem and has its origins in the interpretation of X-rays. In a modern computed tomography (or CT) scanner, the patient lies inside a tube while a sequence of X-ray images are taken from every direction. X-rays are absorbed by dense material, allowing an image to be obtained of our skeleton, our lungs, our brains and other structures within our bodies. The inverse problem for X-rays is to put together all of the images to give a complete image of our internal organs. The mathematical technique behind this process is called a Radon transform, which gives the correct way of integrating a sequence of two-dimensional images in order to construct an accurate three-dimensional picture.

We don't have a Radon transform for social media, but we do – in the form of Equation 5 – have a good understanding of the way social media absorbs and alters social information. In order to reverse-engineer Instagram's process of data deformation, Lina and Michaela used a statistical method called bootstrapping. Each day they took the first 100 messages in their feed and shuffled them randomly to create a new order. They repeated this process 10,000 times, giving a distribution of the orderings that would have occurred if Instagram had just presented each day's posts at random, not prioritizing any particular type of post. By comparing the position of influencers in the actual Instagram feed with these randomized rankings they could identify whether or not influencers had been moved up or down in their feeds.

The results were in stark contrast to those that provoked the #RIP-Instagram outrage. There was no evidence that influencers had been downgraded: their positions in Lina's and Michaela's feeds were not statistically different from what they would have been if the ordering had been created at random. Instagram, they found, was essentially neutral about influencers. It did however strongly prioritize some accounts: friends and family were lifted up to the top of their feed. This promotion of friends was at the expense of news sites, politicians, journalists and organizations generally. Instagram wasn't so much reducing the influence of the influencer, as increasing the influence of friends and family, and pushing accounts which hadn't paid for advertising down their users' feed.

What was most revealing about the #RIPInstagram campaign was the sense of insecurity on the part of the influencers. They suddenly realized that they didn't have as much control over their social positions as they had previously thought. Their positions had been created by an algorithm that promoted popularity, and now they worried it was being taken away by another one that focused on friends.

This research revealed that the real influencers online are not the ones taking pictures of their food and lifestyles. Rather, they are the programmers at Google, Facebook and Instagram who shape the filters through which we see the world. They decide what and who is popular.

For Lina and Michaela, the experiment proved therapeutic. Lina told me how it had changed her perspective of Instagram. She felt she was now making better use of her time on the app. 'Instead of scrolling down trying to find something interesting, I stop after I've seen posts from friends. I know that further down it is just boring stuff,' she told me.

The influencer equation isn't about any one social network. It is about all of them. The equation's power lies in revealing how the structure of online networks shapes the way you see the world. When you are searching for products on Amazon, you are caught in an 'also liked' popularity network as the most popular products are shown to you first. When you are on Twitter, the network combines polarized opinions with a chance to have your views challenged by people all over the world. On Instagram, you are surrounded by friends and family, and sheltered from news and views. Use Equation 5 to look honestly at who and what is influencing you. Write down a connectivity matrix for your social network and see who is in your online world and who is outside it. Think about how this network affects your self-image and how it controls the information you have access to. Move around in it and see how it is affecting the other people you are connected with.

*

In a couple of years Lina and Michaela, who both plan to become maths teachers, will be explaining to teenagers how the algorithms inside their telephone filter their view of the world. For most of the kids, this lesson will help them deal with the complex social network they find themselves embedded in. But a small group of their students will see another possibility – a potential career. They will study hard, understand the maths in more depth and learn how to apply the algorithms used by Google, Instagram and others. A few of these kids might go further still and become part of the rich and powerful elite who control how the information is presented to us.

In 2001, the co-founder of Google, Larry Page, had a patent approved for the use of Equation 5 in Internet searches.[10] The patent was initially owned by Stanford University, where Page was working at the time, and was bought by Google in exchange for 1.8 million shares in the company. Stanford sold the shares for \$336 million in 2005. They would have been worth ten times that amount today. The application of Equation 5 is just one of many patents owned by Google, Facebook and Yahoo for the application of twentieth-century mathematics to the Internet. Graph theory is worth billions to the tech giants who have harnessed it.

That mathematical formulae produced almost one hundred years before these patents were filed can be owned by a university or a company seems to go against the spirit of TEN. The members have always had secrets, but those secrets have usually been shared and used by anyone who wants to learn them. Surely the society should have principles that prevent its members ring-fencing their discoveries or making excessive profits from their carefully collated knowledge?

The answer to that question, it turns out, is far from obvious . . .

6

The Market Equation

$$dX = hdt + f(X)dt + \sigma \cdot \epsilon_t$$

The division of the world into models, data and nonsense gave the members of TEN a sense of certainty. They no longer needed to worry about the consequences; they just needed to put their skills into practice. They turned every problem into numbers and data. They made their assumptions clear. They reasoned rationally and answered the questions posed to them.

Initially the members worked in the civil service and government research agencies. The 1940s and 1950s saw them continue Richard Price's work, developing national insurance schemes for nations and ensuring health services for all. It was at this time that David Cox found himself working in the textile industry, using maths to support industrial growth. The 1960s and 1970s saw the members of TEN taking up positions within research institutes such as Bell Labs in New Jersey, NASA, national defence agencies in countries on both sides of the Cold War, as well as elite universities and strategic think tanks like the RAND Corporation. Vast consolidations of knowledge occurred within these exclusive groups. During the 1980s and 1990s the financial industry caught on, recruiting TEN's members to manage its funds.

Freed from nonsense, the society believed that they alone could solve the world's problems. The rich and powerful agreed, paying them large salaries to manage their investment funds. Governments relied on them to plan for the economic and social future of their countries. Inter-governmental agencies made them central to predicting climate change and setting development goals.

There was, however, something which the TEN mathematicians had forgotten; something A. J. Ayer had made clear in *Language, Truth and Logic*, but which had not been understood to the same degree as the other ideas of the logical positivism that now drove the society forward. When Ayer used the verification principle to separate mathematics and science from nonsense, he found that the non-sense category was much larger than most scientists tended to acknowledge. He showed that morality and ethics were also nonsense.

Ayer proved this point in stages. He started by classifying religious truths. He showed that belief in a God was unverifiable: there was no experiment that could test the existence of the Almighty. He wrote that a believer might claim that God is a mystery that transcends human understanding, or that belief is an act of faith, or that God is an object of mystical intuition. To Ayer these statements were fine, as long as the person of faith was clear that they were non-sense statements. Believers should not and could not imply that God or any other supernatural being had a role in the observable world. The religious convictions of any single individual, or the teachings of any religious prophet, cannot be tested against data and are therefore unverifiable. Or if a religious person contends that their beliefs are subject to verification, then they can be tested against data and (with high probability) proven to be wrong. Religious belief was nonsense.

Up to this point, most of the members of TEN accepted and understood Ayer's argument. It fitted well with their beliefs. They had already rejected miracles and no longer needed a God. But Ayer went further. He handled atheists who argued against religious beliefs with the same calm dismissiveness as he dealt with believers. Atheists were debating nonsense and thus were themselves engaged in the creation of nonsense. The only empirically valid statements on religion involved an analysis of the psychological aspects of the individual believer and/or the role of religion in society. Arguing against religious beliefs was as meaningless as arguing for them.

It didn't stop there. Ayer further rejected the utilitarian argument, made by some of the other members of the Vienna Circle, that we should work towards the greatest happiness for all. He maintained that it is impossible to use science alone to decide what is 'good' or 'virtuous' or to justify an equation about how we should balance happiness now with fulfilment in the future. We can model the rate at which online casinos take money from customers who can't really afford to gamble, but we can't use our model to say whether it is wrong for gamblers to spend their money in any way they want. A climate modeller can state that 'if we don't reduce CO_2 emissions, then future generations will experience an unstable climate and food shortages', but this does not tell us whether it is right or wrong to optimize our life now or to think instead about our grandchildren's well-being. For Ayer, all encouragement to moral behaviour, such as 'we should help others', 'we should act for the greater good', 'we have a moral responsibility to preserve the world for future generations' and 'you shouldn't patent mathematical results' were statements of sentiment, the study area of psychologists, and contained no sensible meaning.

Nor does Ayer's argument permit individualistic sentiments such as 'greed is good' or 'look out for yourself, number one' to be empirically verifiable. Again, these are non-sense, albeit a nonsense embedded deep within our psyche. There is no way to verify such statements against our experience, other than to discuss the relative financial and social success of people following these maxims. We can model the factors that lead a person to become rich or famous. We can measure the personalities of those who succeed. We can talk about how these traits evolved through natural selection. But we can't use maths to prove that certain values are intrinsically good or virtuous. The principle of verification, so powerful in helping the members of TEN to model the world, proved useless in determining its moral path.

If TEN couldn't find any morality from within, then where was its sense of certainty coming from? And, without moral guidance, whose interests was it really serving? Maybe it wasn't quite as righteous as Richard Price had envisaged?

*

I was sitting at a table in one of Hong Kong's finest restaurants, looking out over the harbour. One of the world's largest investment banks had invited me to dine with its best market analysts. Everything was first class, from the flight over with my wife to the five-star hotel and the food we were now eating.

The discussion turned to one of the biggest tensions in their world: the difference in strategy between long- and short-term investments. These men (and one woman) worked mainly on the long-term side of things, managing growth of pension funds. Their decision to invest in a company was decided on its fundamentals, its management structure, its future plans and its position in the market. This was a world they felt they understood and were confident about. If they didn't know what they were doing, then we wouldn't have been sitting in a restaurant with such a spectacular view.

But they were less sure about the short term. Trading had become algorithmic and they didn't understand what the algorithms were doing. The dinner guests asked me which programming languages they should get their new employees to learn? What mathematical skills did they need to have? Which universities had the best Masters degrees in data science?

I tried to answer their questions as best I could, but as I did so I realized that I had missed something very obvious. I had taken something for granted about these people that I shouldn't have. I had assumed, given the spectacular view and the Michelin-star food on my plate, that these guys were like me, that they saw the world through the prism of mathematics, and this was the reason they were now so rich. When, early in the evening, I told them about how I was using the Markov assumption to analyse sequences of possession in football matches, they nodded their heads and looked very knowledgeable. They dropped a few buzz words: machine learning and big data. I realized, of course, that they couldn't know all the details of what I was working on, but I also believed that they had grasped the key ideas.

They didn't want to let down the façade. And now, as they asked about the skills they needed in their recruits, I suddenly saw through it. They had no idea what I was talking about. They knew very little about the equations. They couldn't programme a computer and they saw statistics not as a science, but as a list of numbers to be found in the appendix of annual reports. One of them asked me if calculus was an important skill for a maths graduate.

How could I have been so naive? Why hadn't I noticed before? That afternoon we had listened to some guy who had written a book about why you should 'think slow'. It was all very 'inspirational'. He repeated the word 'slooooow' very slowly, so that we understood that we should wait a long time before making decisions. And he told stories about when he had held a stock a looooong time and it had gone up, or when he had set a biiiiig time interval over which to evaluate his assets. And he told us about a fake beef he had had with some other guy who thought we should do things fast. One example he used to make his point was about an automated trading firm, working out of California. In high-speed trading, even the time it takes for a price to be sent from the West Coast to the trading floor in Chicago takes too long. So this company relocated, taking its main frame nearer to the stock exchange. But the performance of their algorithm dropped. It had worked better when it was further away.

The speaker's conclusion was that the case study supported his conclusion that slow is better. This was clearly incorrect. It was in fact a story about how an algorithm tuned for one time setting might not work at another setting – a trivial detail. At best it could be considered a story about how, if your algorithm is tuned for a different timescale than the others have been using, then you might have an advantage. All the trading algorithms near the stock exchange were tuned to exploit short timescale inefficiencies, while the West Coast traders could exploit inefficiencies on slightly longer timescales. That was until they moved the server. But there was nothing unique about slower timescales.

There certainly exists good-quality research on human decision-making in both economics and psychology, but this speaker did not adhere to basic scientific standards. He gave half-thought-through advice about investments using a false dichotomy about time to appear as if he had a theory. But knocking this one person isn't my point. What bothered me was the way his anecdotes and the stories of other presenters at this conference were lapped up by the attendees: market analysts who knew next to nothing about the algorithms that underpinned their entire business telling each other anecdotes to make them feel clever.

I had let myself become part of it. My role here, it seemed, was to provide more anecdotes of the same type – Premier League betting, scouts at football clubs and Google's algorithms – that would re-affirm my hosts' own belief that they understood how things like high-frequency trading and sports analytics worked. What bothered me more than their lack of knowledge of the technical details of high-frequency trading was that there were genuine lessons to be learnt from the algorithms these traders were using. These were important, practical lessons which could help them take a more balanced approach to their own work. But because they saw the algorithms as a black box, as a way for a small bunch of quants, whose salaries they paid, to somehow engineer profits for them, they didn't seem interested in understanding what it was the quants knew and what they had so far failed to grasp.

More than that, they were scared to ask questions in case they might not be able to understand the answers. I could feel the fear at the table and, to my shame, I pandered to it. Rather than tell them what they needed to learn, I continued to provide the anecdotes they were expecting to hear: I told them about my visit to Barcelona, I told them about Jan and Marius and about how football scouts find new players. They seemed interested and the evening went pleasantly enough. They had lots of anecdotes to tell too, genuinely interesting stories. One of them had recently met Nassim Taleb, a big hero of mine. Another had a daughter studying maths at Harvard. I drank the wine and soaked up the atmosphere. I enjoyed that it was now my turn to tell stories and told them the best way I could.

Don't judge me. There is no reason that I can't enjoy the company of people who don't know the mathematical details of trading. Sometimes they can be even more fun than those who do.

*

Here is what I might have said, if I hadn't been such a hypocrite.

There is no way of unlocking the secret of financial markets without starting with this fundamental equation.

$$dX = hdt + f(X)dt + \sigma \cdot \epsilon_t$$

(Equation 6)

Equations simplify the world by condensing lots of knowledge into a small number of symbols, and the market equation is a brilliant example of this. If we want to unpack the knowledge contained within it, then we need to go through it step by step, very carefully.

The equation describes how X, which represents the 'feeling' of investors about the current value of a stock, changes. The feeling can be either positive or negative, so $X = -100$ is a really bad feeling about the future and $X = 25$ is a pretty good feeling. Economists talk about markets being bullish or bearish. In our model a bullish market is positive $(X > 0)$ about the future and the bear market is negative $(X < 0)$. If we wanted to be more concrete, we might think of our X as the number of bullish people minus the number of bearish people. But at this stage we don't want to get tied down to a particular unit for measuring X. Instead, think of X, loosely, as capturing emotions. Instead of investors, it could be the feeling in a meeting after a redundancy process is announced or after your company has received a big order.

As a convention in mathematics, we tend to put the things we want to explain on the left and the things that we think will do the explaining on the right. And this is exactly what we do in this case. The left-hand side in this instance is dX. The letter d denotes change. So dX means 'the change in feeling'. Notice how the atmosphere in the room dips as you find out your jobs are at risk. This threat of redundancy might be $dX = -12$. Or the newly placed order that is going to sustain your company for the next few years might be $dX = 6$. An even bigger order could be $dX = 15$.

The units, the magnitude of the numbers I use, are not what you should focus on. When we do maths at school, the questions tend to be about adding and subtracting real things, like apples and oranges or money – here we can allow ourselves more freedom. I realize there is no such thing as a $dX = -12$ change in the emotions of your colleagues, but that doesn't mean that we can't write down an equation that tries to capture changes in how a group of people feel. In fact, that is exactly what a share price is: it is how investors feel about the future value of a company. We want to explain changes in our collective feeling about a particular stock investment or the feeling in the office at work, or our feeling about a political candidate, or our feeling about a consumer brand.

The right-hand side of the equation consists of three terms – hdt, $f(X)dt$ and $\sigma \cdot \epsilon_t$ – which we add together. The most important part of each of these terms is the signal, h, the feedback, $f(X)$, and the standard deviation or noise, σ. The factors which multiply these terms indicate that we are interested in changes (d) in time (t). The noise is multiplied by ϵ_t which denotes little random bumps in time. These terms model our feelings as driven by a combination of signal, social feedback and noise. Now we are close to grasping something fundamental, let's make the equation more concrete through an example.

*

You might be wondering if you can use the market equation to choose your pension scheme. I'm afraid you will have to wait for an answer. There are more important questions at hand, such as whether you should go and see the new Marvel film. Or what type of headphones you should buy. Or where you are going to go on holiday next year.

Let's take the decision to buy new headphones. You have saved £200 to get a nice pair and now you are browsing online looking for the best buy. You go to Sony's webpage and read the technical specs; you read the reviews of the Japanese brand Audio-Technica; you see all the celebrities and sports stars wearing Beats. Which should you choose?

I can't tell you which headphones to buy, but I can tell you how to think about the problem. Questions like this are all about separating the signal, h, the feedback, $f(X)$, and the noise, σ. Let's start with Sony, using a variable X_{Sony} to measure how much consumers like Sony as a brand. My first-ever good-quality Walkman and pair of headphones, bought in 1989, second-hand from Richard Blake, were Sony. They are classic and dependable. In terms of Equation 6, Sony has a fixed value of $h = 2$ and take the time unit to be $dt = 1$ year. Since units of 'feeling' are arbitrary, the value 2 isn't in itself important. What is important is the size of the signal relative to the social feedback and noise. For Sony, we choose $f(X) = 0$ and $\sigma = 0$. There is only signal.

If we start with $X_{\text{Sony}} = 0$ in 2015, then, because $dX_{\text{Sony}} = h \cdot dt = 2$, in 2016 we have $X_{\text{Sony}} = 2$. In 2017 we have $X_{\text{Sony}} = 4$ and so on, until in 2020, $X_{\text{Sony}} = 10$. The positive feeling about Sony grows because of a positive signal.

You know much less about another brand, Audio-Technica. They have had good reviews on a couple of YouTube channels. The sound geek in your local hi-fi shop claims they are the hottest thing among Japanese DJs, but you don't have a lot of information to go on. There is a risk in taking advice from only one or two sources and it is this risk which makes the noise. Since recommendations for the Japanese DJ headphones come from only a small number of places, we give them $\sigma = 4$, so the noise is twice the size of the signal.

The market equation for Audio-Technica is $dX_{\text{AT}} = 2dt + 4\epsilon_t$. We can think of the last term, ϵ_t, as producing a random number each year. Sometimes this will be positive, sometimes negative, but on average ϵ_t will be zero and have a variance equal to one.

By picking random values for ϵ_t we can simulate the random nature of the information you receive about Audio-Technica. This is what is typically done by financial quants modelling changes in share prices. For any given problem they run millions of simulations and look at a distribution of outcomes.

Here, to illustrate how these simulations work, I am going to 'run' one simulation, making up the random values as I go. Let's imagine that, for 2015, the random value is $\epsilon_t = -0.25$. If so, then $dX_{\text{AT}} = 2 - 4 \cdot 0.25 = 1$. If the next year $\epsilon_t = 0.75$, then $dX_{\text{AT}} = 2 + 4 \cdot 0.75 = 5$, and if in 2017 $\epsilon_t = -1.25$, then $dX_{\text{AT}} = 2 - 4 \cdot 1.25 = -3$. Your confidence in Audio-Technica grows over time (X_{AT} is 1, 6 and 3 in 2016, 2017 and 2018, respectively) but much more erratically than for Sony.

Finally, we have our social feedback product: Dr Dre's Beats. These headphones are about sharing how you look on social media and creating a feeling that drags others along. Beats let you believe the hype. As celebrities and online influencers get into the feeling, more people adopt them and the feeling grows even stronger. In terms of our model, we could, for example, set $f(X) = X$ so that the feeling for Beats grows in proportion to itself. The more love there is for Beats, the more love is generated. This gives a market equation $dX_{BEATS} = 2dt + X_{BEATS}dt + 4 \epsilon_t$: 2 units of growth, X_{BEATS} of social feedback and 4 units of noise underlie the growth of Beats.

Let's imagine that Beats starts with a bad year in 2015 with a random bump of $\epsilon_t = -1$. Assuming that initially we set $X_{BEATS} = 0$, then, when we apply the market equation, we get $dX_{BEATS} = 2 + 0 - 4 \cdot 1 = -2$. By the start of 2016 our confidence in Beats is negative, $X_{BEATS} = -2$. The next year goes a bit better in terms of noise, $\epsilon_t = 0.25$, but the social feedback limits the improvement. $dX_{BEATS} = 2 - 2 + 4 \cdot 0.25 = 1$. So $X_{BEATS} = -1$ in 2017. During 2018, $\epsilon_t = 1$, and the feeling about Beats starts to grow, $dX_{BEATS} = 2 - 1 + 4 = 5$. Now $X_{BEATS} = 4$ the social feedback takes off and despite a poorer year in 2019, $\epsilon_t = 0.0$, we have $dX_{BEATS} = 2 + 4 + 0 = 6$ and feeling about Beats, $X_{BEATS} = 10$, carries it through. The social feedback term amplifies both the bad and the good: it can make it harder for a product to take off in the first place, but once a feeling is established then that feeling grows even stronger.

I am, of course, creating a caricature of Sony, Beats and the Japanese DJ brand. So, before any of them sue me, I want to point out the genuine difficulty for you as a consumer. By browsing online and even by asking your friends, what you are measuring are the feelings people express about different headphones. For our simulation, these are summarized in Figure 6.

Over the years the best-quality product changes, according to consumer feeling. In 2016 and 2018 Sony's is considered the best. In 2017 it is Audio-Technica. And in 2019 and 2020, Beats appears to be the best.

You might be tempted to conclude from my description above that you should go for Sony, which has the most reliable signal. But remember, all the other products also have a true signal and, in this case, $b = 2$ is the same for every product. What the right-hand side of the market equation tells you to do is to dig deeper. For any given brand, the equation will be driven by a combination of all three factors. The challenge for you, as a listener, is to find that signal within the feedback and the noise. The same reasoning applies for all other consumer products, from the latest blockbuster or online game to sneakers and handbags. What you are most often subjected to is a

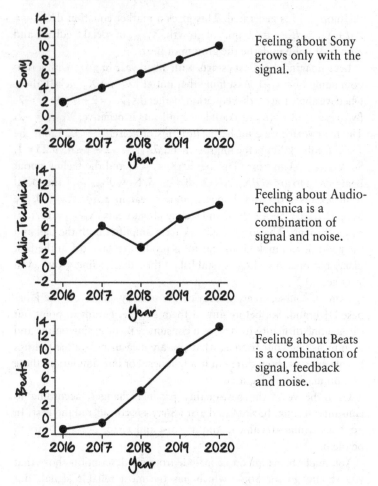

Feeling about Sony grows only with the signal.

Feeling about Audio-Technica is a combination of signal and noise.

Feeling about Beats is a combination of signal, feedback and noise.

Figure 6: How the feeling for the three products changes over time

feeling about a product, but what you really want to know is the underlying quality.

For stock markets the problem is the same. We often only know the growth of a share price, dX, but what we really want to know is the strength of the signal. Is there a lot of social feedback, creating hype around a product? What are the sources of confusing noise?

For centuries, the members of TEN saw only the signal. Inspired by the inescapable pull of Newtonian gravity, the eighteenth-century Scottish economist Adam Smith described an invisible hand that leads the market towards equilibrium. Barter and exchange of goods balanced supply with demand. Italian engineer Vilfredo Pareto formalized Smith's view in calculus, describing our continual economic evolution towards optimality. The signal of profit would lead us deterministically to stable prosperity. Or so they believed.

The first signs of instability – the Dutch tulip mania and Britain's South Sea Bubble – were few and far between, giving little real cause for concern. It was only when capitalism spread over the whole globe that the booms and busts required some explanation. From the Great Depression of 1929 to the stock market crash of 1987, repeated crises showed society that markets were not perfect – they could be messy and wildly fluctuating. The noise became just as strong as the signal.

As physics evolved at the start of the twentieth century, from Newton to Einstein, so too did the mathematics of the markets. Even before he published his theory of relativity, Einstein had explained how the movement of pollen in water was caused by a random bombardment of water molecules. The way in which our economic prosperity was buffeted by external events appeared to be perfectly captured by this new mathematics of randomness, and TEN's members started to formulate a new theory. In 1900, Louis Bachelier, a French mathematician, published his thesis *The Theory of Speculation*, setting out the first two components of Equation 6. For most of the twentieth century, noise became a new source of profit. Extensions to the basic theory, such as the Black–Scholes equation, are used to devise and price derivatives, futures, puts and options. The society's members were recruited both to create and to control these new financial models. In effect, they were put in charge of the world's money supply.

Just as Newton's deterministic calculus was the wrong model for financial markets, so too this view of the markets as bombarded by noise was missing a vital element: us, the market participants. We are not particles, moved only by events, but active agents, both rational and emotional. We search for the signal in the noise, and, as we do so, we influence each other, we learn from each other and we manipulate each other. Human complexity cannot be ignored by mathematical theory.

Spurred on by this revelation, some of the members of TEN took a new direction. The Santa Fe Institute in New Mexico brought together mathematicians, physicists and scientists from all over the world. There they started to sketch out a new theory of complexity, one which tried to account for our social interactions. The theory predicted large, unpredictable fluctuations in share prices, brought on by herding by traders. As volatility increased, the models said, we should expect even more spectacular booms and busts than we have seen in the past. The researchers published warning after warning in high-profile scientific journals.[1] TEN's secrets were, as always, published in plain sight, for all to read. Unfortunately very few bothered to do so.

One of the Santa Fe researchers, J. Doyne Farmer, left the Institute to put the ideas into practice. He told me later that it had been hard work, harder than he had ever imagined, but it had paid off. Through the Asian crisis of 1998, the dotcom crash of 2000 and then the financial crisis in 2007, Farmer's own investments were safe, insured against the turmoil that brought down financial institutions and governments, and that sowed the seeds of political discontent across Europe and the USA.

The mathematicians could, with some justification, tell us that they had known all along that the crashes were on their way. They were prepared when others were not. While many lost out, the members of TEN were still collecting their profits.

*

I am getting ahead of myself. The historical account of how mathematics went from understanding markets in terms of a simple signal, to appreciating the noise and ultimately to embracing social feedback, is fine, but it misses an important point. It sounds too much like a tale of how human follies at one stage were improved by a new way of thinking.

It is true that mathematicians have learnt from their mistakes over the last century and stayed ahead of everyone else, but there is another important point that needs to be made: mathematicians really have no idea how to find the true signal in the financial market's noise.

That's quite a strong claim on my part, so I need to explain it step by step. The secret to the initial successes of the market equation can be found in Chapter 3. The mathematicians working in finance in the 1980s could separate the signal from the noise by collecting enough observations. Louis Bachelier's first version of the market equation didn't contain $f(X)$: it simply described the growth of confidence in a company, and thus its share price, as a combination of signal and noise. That mathematical knowledge alone allowed traders to reduce the randomness their clients were exposed to. It gave them an edge over those who didn't understand the randomness, who were confusing signal and noise.

For over a decade before the financial crisis in 2007 a group of theoretical physicists had argued that a market equation based purely on signal and noise was dangerous. They showed that the model didn't produce enough variation in share prices to explain the massive booms and busts observed on the stock exchange during the previous century. The tech bubble and Asian crisis of 1999 both involved share prices crashing to values that a simple signal and noise model would never have predicted.

To understand the scale of these large deviations, think back to de Moivre and his coin-tossing trick. He found that the number of heads after n coin tosses usually fell within an interval of size proportional to the $\sqrt{n}$. The Central Limit Theorem (CLT) extended de Moivre's result to say that the same $\sqrt{n}$ rule applies for all games and even in many real-life situations, such as opinion polls. The key assumption for the CLT is that events are independent. We add up the outcome of independent spins of the roulette wheel or ask the independent opinions of lots of different people.

The simple signal and noise market model also assumes independence in price changes. Under the model, future values of shares should thus follow the $\sigma\sqrt{n}$ rule and the Normal curve. In reality, they don't. Instead, as the theoretical physicists in Santa Fe and elsewhere showed, the variation in future share prices can become proportional to higher powers of n, such as $n^{2/3}$ or even proportional to n itself.[2] This makes markets scarily volatile. And it makes predictions almost impossible: a share can lose its entire value in one day, as if de Moivre tossed a coin and came up with 1,800 tails.

The reason for these large fluctuations is that traders don't act independently from each other. In roulette, one spin of the wheel doesn't depend on the last and the Central Limit Theorem applies. But on the stock market, one trader who sells causes another to lose confidence and sell too. This invalidates the assumptions of de Moivre's Central Limit Theorem and the fluctuations in share prices are no longer small and predictable. Stock market traders are herd animals, following each other into one boom and bust after another.

Not all the financial mathematicians understood that the Central Limit Theorem didn't apply to markets. When I met J. Doyne Farmer in 2009, he told me about a colleague at one trading firm, which unlike Farmer's own company had lost a lot of money during the 2007–2008 crisis, who referred to the Lehman Brothers investment bank crash as a 'twelve-sigma event'. As we saw in Chapter 3, 1-sigma events occur one time in three, 2-sigma events occur about one time in twenty and a 5-sigma event about one time in 3.5 million. A 12-sigma event occurs 1 time in, well, I'm not sure, actually, because my calculator fails when I try to find anything larger than a 9-sigma. It is very unlikely, in any case, and there was no way an event like this should occur unless the model is very wrong.

The theoretical physicists may have uncovered the maths behind large deviations, but they certainly weren't alone in describing the herd mentality of traders. Nassim Nicholas Taleb's two books, *Fooled by Randomness* and *The Black Swan*, provide an amusingly arrogant, but superbly prescient analysis of the pre-2007 financial world. Robert J. Shiller's book from the same period, *Irrational Exuberance*, gives a more academic and thorough treatment of similar ideas.[3] When theoretical physicists, hard-nosed quantitative investors and Yale economists all provide the same warnings about a model, then it is probably a good idea to pay attention.

At the turn of the millennium, many theoretical physicists who went to work in the finance industry found an edge in the market. They kept that edge throughout the financial crisis and won big when the markets fell. By adding the $f(X)$ term to their market equation, they were ready for events such as the collapse of Lehman Brothers, when traders had followed each other into extreme and risky positions.

All financial mathematicians now know that markets are a combination of signal, noise and herding: their models show that the crashes will occur and allow them to get a good idea of how large they will be in the long run. But the financial mathematicians don't know when or why these crashes take place, at least no more than that it has something to do with a herd mentality. They don't understand the fundamental reasons for the ups and downs. When I was at the online casino in Chapter 3, I knew that the game was biased in favour of the casino. I knew that the signal was a 1/37 average loss per spin. I could work this out by looking at the structure of the roulette wheel. In Chapter 4, Luke Bornn looked at the contributions of each basketball player to the overall team performance in order to measure skill. They found the skill signal by combining their knowledge of the game with well-chosen assumptions. In Chapter 5, when Lina and Michaela reverse-engineered the Instagram algorithm, they could start to see how their world view was distorted by social media. In all of these examples the model gives insight into how roulette, basketball and social media, respectively, operate. The market equation does not, in itself, provide understanding.

Researchers have, at various times, tried to see if they can go one step further and find the true signal in markets. In 1988, after the 1987 Black Wednesday crash, David Cutler, James Poterba and Larry Summers at the National Bureau of Economic Research wrote a paper entitled 'What moves stock prices?'[4] They found that factors such as industrial production, interest rates and dividends that affected stock market returns only explained about one third of the variation seen in the value of the stock market. Then they looked to see if big news events, such as wars or changes in the Presidency, played a role. Big news days did see significant changes in share values, but there were also a fair number of days on which market movements were large when there was no news. The vast majority of stock exchange movements can't be explained by external factors.

In 2007, Paul Tetlock, Professor of Economics at Columbia University, created a 'pessimism media factor' for the 'Abreast of the Market' column of the *Wall Street Journal*, which is written directly after trading closes each day.[5] The factor measured the number of times different words were used in the column and thus the overall sentiment of the writer's report about the day's trading. Tetlock found that pessimistic words were associated with falls in shares the next day, but these drops were reversed later in the week. He concluded that the 'Abreast of the Market' column was unlikely to contain any useful information about longer-term trends. Other studies have shown that gossip in Internet chat rooms and even what people say to each other on trading floors can predict the amount of trading but not the direction of movement of the market.[6] There are simply no reliable rules for predicting the future value of shares.

I want to be clear about two things here. First, these results don't mean that news about a company doesn't affect its share value. Facebook shares dropped in value after the Cambridge Analytica scandal. BP shares fell after the Deepwater Horizon oil-spill disaster. In these cases, though, the events that caused the share price changes were even less predictable than the share value itself, making them more or less useless to the profit-seeking investor. Once you have heard the news, then so too has everyone else. The opportunity for an edge is gone.

Secondly, I want to re-emphasize that models based on the market equation do provide useful long-term planning for risk. A mathematician friend of mine, Maja, works for a high street bank. She uses Equation 6 to assess the various risks which the bank is exposed to, then buys insurance to protect it against inevitable ups and downs. Maja finds that non-mathematicians seldom understand the limitations of the models she uses. Last time we met for lunch, together with her colleague Peyman, she told me that 'The biggest problem I see among non-mathematicians is that they take the results of models literally.'

Peyman agreed. 'You show them a confidence interval for some time in the future and they take that as true. Very few of them understand that our model is based on some very weak assumptions.'

What Maja and Peyman struggle with is the perception that because it is maths, it must be true. The market equation isn't like that. Its main message is that we have to be careful because almost anything could happen in the future.

This view of financial markets, where we can insure ourselves against fluctuations in the market but we can't understand why they have occurred, is shared by many traders. When the markets temporarily melted down and bounced again at the start of 2018, Manoj Narang, CEO of quantitative trading firm MANA Partners, told the business news organization Quartz that 'Understanding why something happened in the market is only slightly easier than understanding the meaning of life. A lot of people have educated guesses, but they don't know.'[7]

If the traders, bankers, mathematicians and economists don't understand the reasons markets move, then what makes you think that you do? What makes you think that Amazon shares have reached their peak or Facebook shares will continue to fall? What makes you so confident when you talk about getting into the housing market at the right time?

In the summer of 2018 I was a guest on CNBC's *Power Lunch*, one of the biggest business news shows in the USA. I have been in news studios before, but this was on a whole different scale: a massive open hall, the size of an ice hockey stadium, filled with journalists running backwards and forwards between desks. The screens were everywhere, showing video feeds of shiny offices in Seattle, high-speed underground computer halls in Scandinavia, large factory complexes in China and images of a business meeting in an African capital city. My hosts took me into the editing room to watch as this information was cut together to produce a live stream. The scenes from around the world were overlaid with scrolling share price numbers and breaking news headlines.

What the market equation has taught me is that almost everything on those screens is meaningless noise or social feedback. It is nonsense. There is nothing useful that can be gained by watching daily changes in share price updates or professional pundits explaining why you should or shouldn't buy gold. There are many investors, including some of those I met in Hong Kong, who can identify good investments by thoroughly researching the fundamentals of the businesses they invest in. But, with the exception of a systematic investigation of how a company is run and works internally, all investment advice is random noise. That includes the motivational musings of gurus who happen to have made money in the past.

This inability to predict the future based on the past applies to our personal finances. If you are buying a house, don't worry about how the prices in that area have changed over the last few years. You can't use the trend to predict the future. Instead, you need to be acutely aware that house prices experience massive fluctuations, both upwards and downwards, as market feeling changes. Make sure you are mentally and financially prepared for both eventualities. Once you have prepared yourself, then buy the house you like best and which you can afford. Find a neighbourhood that you like. Decide how much time you are willing to invest renovating the property. Look at travel time to work and school. It is the basics, the market fundamentals, which are important, not whether your house is in an 'up and coming area' or not.

When it comes to buying shares, don't overthink it. Find companies you believe in, invest and see what happens. In addition, put some money in an index-linked investment fund, which spreads your investment over shares in a large number of companies. Make sure you have a good pension. You can't do more than that. Don't stress it.

To test the true quality of three pairs of headphones, the answer is simple. Make a playlist of your favourite ten songs, and listen to them on each pair of headphones one at a time. Randomize the order you listen to each song on each pair of headphones. Then rank the sound. Don't listen to your friends or the online reviews. Listen to the signal.

*

Mathematicians are sneaky. Just after we tell you that everything is random, we find a new and different edge. When we find it isn't possible to predict long-term trends in share prices using maths, then we move in the opposite direction. We look instead at shorter and shorter timescales. We find an edge in places that humans can't possibly make calculations.

On 15 April 2015, Virtu Financial launched itself on the stock market. The company, founded seven years earlier by financial traders Vincent Viola and Douglas Cifu, had developed novel methods for high-frequency trading: buying and selling shares within milliseconds of a trade being performed at a stock exchange on the other side of the country. Up until the point of its share offering, Virtu had been extremely secretive about its methods and how much money it was making. But in order to float on the stock exchange, through an IPO (initial public offering), it needed to open up its finances and details of its business for inspection.

The secret was out. During five years of trading, Virtu made a loss on only one single day. By any standards this result was astounding. Financial traders are used to dealing with randomness; they have learnt to deal with weeks or months of losses as an inevitable part of eventually making profits. Virtu had eliminated the downs from trading and was on the up.

Virtu's initial stock exchange evaluation was $3 billion.

Intrigued by the guaranteed daily profits, Greg Laughlin, Professor of Astronomy at Yale University, wanted to work out how Virtu's performance could have been so reliable.[8] Douglas Cifu had admitted to Bloomberg that only '51 to 52%' of Virtu trades were profitable.[9] This statement puzzled Laughlin initially: if 48% to 49% of individual trades made a loss, then it would take a very large number of trades to make a guaranteed daily profit.

Laughlin looked in more detail at the type of trades made by Virtu. The company made its profits by knowing price changes in advance of its competitors. Its IPO documents revealed that the company controlled a corporation, Blueline Communications LLC, who had developed microwave communication technology capable of sending pricing information between the stock exchanges in Illinois and New Jersey within approximately 4.7 milliseconds. In his 2014 book about high-frequency trading, *Flash Boys*, Michael Lewis had found that the fibre-optic pathway between the two exchanges had a latency of approximately 6.65 milliseconds. Virtu had an approximately 2-millisecond edge over those using fibre.

At a timescale of 1 or 2 milliseconds, the profit margin is around $0.01 per traded share. The margins meant that often trades would occur where no profit or loss was made; the trade would be a scratch. Assuming that 24% of trades were loss-making and 25% were scratches, Greg Laughlin worked out the average profit per trade to be $0.51 \cdot 0.01 - 0.24 \cdot 0.01 = \$0.0027$ per share. Given Virtu's reported traded income of \$440,000 per day, this implied that Virtu was making 160 million share trades per day.[10] That was between 3% and 5% of the total US equity market. They were taking a tiny cut from a sizeable proportion of all available trades. The smallest of edges at the highest possible speeds was giving large, guaranteed profits.

I contacted both Vincent Viola and Douglas Cifu and asked for an interview. Neither of them got back to me. So I rang my friend Mark,[11] a mathematician working for another big quantitative trader, and asked him if he could fill me in on how companies like Virtu worked. He outlined five different ways in which high-frequency traders found their edge. The first was through speed, by having the fastest communication channels, like the microwave technology developed by Blueline. Traders could always be sure of knowing the direction of trades before their competitors. The second was through computing power. Loading a trading calculation into a computer's central processor unit takes time, so teams of up to a hundred developers utilize the graphics cards inside their machines to process trades as they come in.

The third edge, and the one utilized most often by Mark and his own team, builds on Equation 6. A popular form of investment over recent years has been exchange-traded funds (ETF), which are 'baskets' of investments in different companies in a larger market such as the S&P 500 (a list that measures the stock performance of the 500 largest companies in the USA). Mark explained to me, 'We look for arbitrages between the individual share values in the ETF and the ETF itself.' An arbitrage is an opportunity to make risk-free money by exploiting price differences in the same commodity. If, for a sufficient number of milliseconds, the individual values of all the shares in an ETF don't reflect the value of the ETF itself, Mark's algorithms can set up a sequence of buys and sells which make a profit from these price differences. Mark's team identifies arbitrages, not only in the current share price, but also in future prices. A variation of the market equation is used to value options to buy or sell a share one week, one month and one year into the future. If Mark and his team can calculate the future value of both the ETF and all the individual shares before anyone else, then they can make a risk-free profit.

The fourth edge comes from being a big player. 'The more you trade, the cheaper the transactions become,' Mark explained. 'Another advantage is already having cash or stock loans which can be used to cover investments that take three or four months to pay out.' Basically, the rich get richer because their capital is bigger and their costs lower.

The fifth way of finding an edge was one that Mark had never used himself in fifteen years of trading at the highest levels with millions of dollars of capital: it consisted of trying to predict the true value of the shares and commodities being traded. There are traders who look at the fundamentals of different businesses, using experience and good judgement to make investment decisions. Mark isn't one of them. 'I take the position that the market is smarter than me about the prices, then I look at whether the futures or options are priced properly, assuming the market is correct.'

Mark had returned to what I believe is the most important lesson from the market equation, a lesson which applies not only to our economic investments, but also to investments in friendships, in relationships, in work and in our free time. Don't believe that you can reliably predict what will happen in life. Instead, make decisions that make sense to you, which you truly believe in (here you should use the judgement equation, of course). Then use the three terms in the market equation to prepare yourself mentally for an uncertain future. Remember the noise term: there will be many ups and downs which lie outside your control. Remember the social term: don't get caught up in the hype or discouraged when the herd doesn't share your beliefs. And remember the signal term: that the true value of your investment is there, even though you can't always see it.

*

TEN has controlled randomness with increasing confidence over the last 300 years, taking money from the investors who don't know the code. Those who don't know the mathematical secrets see a stock rise, believe that there was an underlying signal, and invest. They see a stock fall and they sell. Or they do the opposite, trying to second-guess the market. In neither case did they consider the possibility that they were guided primarily by noise and feedback.

Outsiders' understanding of the finance game has slowly improved. Members of TEN listen patiently while part-time gamblers and amateur investors talk to them about the signal and the noise. Phrases like 'fooled by randomness', 'finding the signal', 'signal to noise ratio' and 'two sigma' are widely used – reeled off freely and with apparent confidence. While this talk continues, TEN keeps finding new edges on shorter and shorter timescales without ever looking for the signal. Their algorithms take advantage of arbitrages of almost every trade.

Greg Laughlin looked more closely at Virtu's trades after reading Michael Lewis's book, *Flash Boys*, and an article in the *New York Times* by American economist Paul Krugman on the subject.[12] By email Greg told me that 'The idea [expressed in Krugman's article] is that high frequency traders use sophisticated, morally suspect methods to unfairly extract money from the market.' Yet the data from Virtu simply didn't support that view: the company was taking less than 1% per trade to improve the overall efficiency of the market. 'If one has a legitimate reason to buy a stock, namely for long-term gain and based on sound economic fundamentals, then transaction costs are extremely low,' Greg told me. 'If one tries to beat the market by day trading, or if one panics and wants to unload a portfolio during a moment of high volatility, then high frequency trading will take advantage of those behaviours.'

When day traders played the stock market in the same way as amateur gamblers bet on sports, the mathematicians were there to exploit the traders' lack of understanding of randomness. As always, TEN made its money through tiny edges, anonymously and invisibly, and without a fuss.

*

The question of morality was the last thing I asked Mark about. How did he feel making high-speed profits from other people's transactions? I put it to him that, when his team finds an arbitrage, his profits come from the pension funds and the investors who are not trading as fast and as accurately as he is. I asked him how he felt about taking money from my own and others' pension investments?

We were talking over the phone, Mark standing in his garden in a leafy suburb of a major European city. I could hear the birds singing in the background as he thought carefully about how to answer. And I felt acutely embarrassed about asking him a question that I knew lay outside of the technical aspects of his job, a question about his own contribution to society. A person like Mark – who has made his money anonymously, without a fuss, applying the equations again and again – is deeply honest by nature. He is forced to analyse his own contribution with the same rigour that he analyses the stock market, the same careful way he analyses everything. I knew whatever he said would be factually correct.

'Instead of talking about the morality of a single trade, what I ask myself is whether the markets are more or less efficient because of my trading? Does your pension fund pay higher or lower costs overall?' he said. 'Before high-speed trading started, if you called up your broker and asked them for a sell price and a buy price, the difference between those two prices was much larger than it is now.'

Mark described somewhat dubious practices where brokers would take quite large percentage cuts for making transactions. 'Now there are a much smaller number of much more sophisticated firms making a tiny fraction from every trade.' Old-school brokers, without the ability to properly calculate futures and who took larger cuts, have gone out of business. 'So my impression is that the markets are more efficient than they used to be, but I can't be sure, because the volume of trading has also increased,' he said. He admitted that he didn't have all the numbers so couldn't say too much more, but what he told me was in line with what I had also heard from Greg Laughlin.

Mark's answer to the role of high-frequency trading wasn't definitive, but it was honest. He didn't use self-justification, excuses, ideology or pithy arguments. He turned a moral question into a financial one. It was an answer A. J. Ayer would have approved of. It was the answer of a member of TEN: neutral and without nonsense.

7

The Advertising Equation

$$r_{x,y} = \frac{\sum_i (M_{i,x} - \overline{M_x})(M_{i,y} - \overline{M_y})}{\sqrt{\sum_i (M_{i,x} - \overline{M_x})^2 \sum_i (M_{i,y} - \overline{M_y})^2}}$$

At first I thought the email was spam. It began with the greeting 'Mr Sumpter:' – and there aren't many real people who use a colon to open an email. Even as I read the text, a friendly request from the US Senate Committee on Commerce, Science and Transportation in Washington DC to conduct an interview with me, I remained sceptical. What felt strange was the very fact it came in the form of an email at all. I'm not sure how I expected a Senate Committee to contact me, but I was suspicious of the juxtaposition of a long and detailed committee title and the informal request for help. It didn't sit right.

It was right, though. The Senate Committee did want to talk to me. I replied briefly and positively, and a few days later I was on a Skype call to the staff of the Republican side of the committee. They wanted to know about Cambridge Analytica, the company that Donald Trump had hired in order to reach out to voters on social media, and which had allegedly collected data from tens of millions of Facebook users. In the media, there were already two sides to the Cambridge Analytica story. One was the slick presentation of Alexander Nix, the then CEO, who claimed to be using algorithms in political campaigns by micro-targeting voters' personalities. The other side was the colourfully haired whistleblower, Chris Wylie, who claimed that he had helped Nix and his company to create a tool for 'psychological warfare'. Subsequently, Wylie regretted the way he had acted to help get Trump elected, while Nix had been building his business in Africa on the back of his 'success'.

I had researched the algorithm used by Cambridge Analytica in detail during 2017, the year before the scandal broke, and had reached a conclusion that contradicted both Nix and Wylie's versions of events. I doubted that the company could have influenced the US presidential election. They had certainly tried, but I found that the methods they claimed to have used to target voters were flawed.[1] My conclusions had put me in the strange situation of questioning both of the current narratives. And this was why the Senate Committee wanted to talk to me. Most of all, the Republicans in the Trump administration in the spring of 2018 wanted to know what to make of the massive scandal that had erupted about advertising on social media.

*

Before we can help the Senate members, we first need to understand how social media companies view us. To do that we are going to treat people as data points (as they do), starting with one of the most active and important data points of them all: teenagers. This group wants to see as much as possible as quickly as possible. Every evening they can be found – either in groups on a couch or, increasingly, alone in their bedrooms – rapidly clicking and swiping on their favoured social media platforms, Snapchat and Instagram. Through the small window provided by their phones they are able to see an incredible view of the world: dwarves falling off skateboards, couples going on truth-or-dare Jenga dates, dogs playing the game *Fortnite*, small kids pushing their hands slowly into Play-Doh, teenage girls destroying their make-up or 'hooked' stories, which consist of texted dialogue between imaginary college students. These features are interspersed with celebrity gossip, a very occasional piece of real news and, of course, regular and unending advertisements.

Inside Instagram, Snapchat and Facebook a matrix is generated about our interests. This matrix is a spreadsheet-like array of numbers where the rows of the matrix are people and the columns are the types of 'posts' or 'snaps' they are clicking on. In mathematics we represent this spreadsheet of teenage snapping in a matrix denoted M. Here is an example to illustrate – on a much smaller scale – what a social media matrix looks like for twelve users.

Food	Make-up	Kylie Jenner	PewDiePie	Fortnite	Drake	
8	6	6	0	0	2	Madison
1	6	1	4	0	9	Tyler
2	0	0	9	5	3	Jacob
5	0	9	8	7	2	Ryan
5	9	7	1	0	1	Alyssa
3	6	9	1	2	3	Ashley
5	7	7	1	2	4	Kayla
6	3	3	5	6	9	Morgan
6	0	0	0	2	8	Matt
1	4	9	8	2	1	Jose
8	7	8	2	3	1	Sam
2	0	1	8	7	4	Lauren

$M = $ (to the left of the matrix above)

Each entry of M is the number of times that one of the teenagers has clicked on a particular type of post. So, for example, Madison has watched 8 food posts, 6 posts each about make-up and the celebrity Kylie Jenner, no posts on YouTuber PewDiePie or on the video game *Fortnite* and 2 posts about the rapper Drake.

We can already get quite a good idea, just from looking at this matrix, what sort of person Madison is. Feel free to picture her in your head and then take a few seconds to imagine some of the other characters I have created here, using the snaps they look at as a guide. Don't worry. They aren't real people. You can be as judgemental as you like.

There are a few others in the matrix who are similar to Madison. Sam, for example, loves make-up, Kylie Jenner and food, but has only a passing interest in the other categories. There are also people who are very different than Madison. Jacob loves PewDiePie and *Fortnite* above everything else, as does Lauren. Some don't quite fit into these two stereotypes. Tyler, for example, likes Drake and make-up, but is also fairly interested in PewDiePie.

The advertising equation is a mathematical way of automatically stereotyping people. It takes the following form:

$$r_{x,y} = \frac{\sum_i (M_{i,x} - \overline{M_x})(M_{i,y} - \overline{M_y})}{\sqrt{\sum_i (M_{i,x} - \overline{M_x})^2 \, \sum_i (M_{i,y} - \overline{M_y})^2}}$$

(Equation 7)

It measures the correlation between different categories of snaps. For example, if people who typically like Kylie Jenner also like make-up, then $r_{make-up,Kylie}$ will be a positive number. In this case, we would say there is a positive correlation between Kylie and make-up. But, if people who like Kylie tend not to like PewDiePie, then $r_{PewDiePie,Kylie}$ will be a negative number and we call it a negative correlation.

In order to understand how Equation 7 works, let's break it down step by step, starting with $M_{i,x}$. This is the value in row i, column x of the matrix M. So $M_{Madison,make-up} = 6$, because Madison watched 6 posts about make-up. The row is i = Madison and the column is x = make-up. In general, wherever we see $M_{i,x}$ we simply look up entry i,x in our matrix M. Now look at $\overline{M_x}$. This denotes the average number of posts watched per user about category x. For example, the average number of posts watched about make-up across all our teenagers was $\overline{M_{make-up}} = 4$, i.e. $(6 + 6 + 0 + 0 + 9 + 6 + 7 + 3 + 0 + 4 + 7 + 0)/12 = 4$.

If we subtract the average interest in make-up from the number of posts watched by Madison we get $M_{i,x} - \overline{M_x} = 6 - 4 = 2$. This tells us that Madison has an above average interest in make-up. Similarly, by calculating $\overline{M_{Kylie}} = 5$ we see that she also has a (slightly) above average interest in Kylie, i.e. $M_{i,y} - \overline{M_y} = 6 - 5 = 1$, for i = Madison and y = Kylie.

The last two paragraphs are a form of bookkeeping, or, as we say in mathematics, notation. They set things up for the really powerful, interesting idea at the heart of Equation 7; that is where we multiply $(M_{i,x} - \overline{M_x}) \cdot (M_{i,y} - \overline{M_y})$ to find things people tend to have in common. For Madison, we have:

$$(M_{Madison,make-up} - \overline{M_{make-up}}) \cdot (M_{Madison,Kylie} - \overline{M_{Kylie}}) = (6 - 4) \cdot (6 - 5) = 2 \cdot 1 = 2$$

revealing a positive relationship between her interest in Kylie and make-up.

For Tyler, the relationship between make-up and Kylie is negative, $(6-4) \cdot (1-5) = 2 \cdot (-4) = -8$, because he is only interested in the former. For Jacob it is positive, $(0-4) \cdot (0-5) = (-4) \cdot (-5) = 20$, because he doesn't like either (see Figure 7). Notice a subtlety here. Both Jacob and Madison have a positive value, even though they have opposite feelings about Kylie and make-up. Both their views suggest that Kylie and make-up are correlated, even though Jacob never looks at either. Tyler's use of social media doesn't fit the pattern.

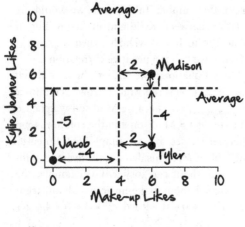

For each individual the distance between the average number of likes for everyone and their number of likes is measured for both make-up and Kylie.

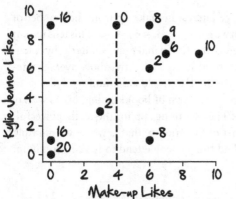

The correlation is the sum of the two distances multiplied together. So Madison contributes 2.1 = 2, Tyler $2 \cdot -4 = -8$ and so on.

In this example only two people differ in their tendency to 'like' Kylie and make-up.

Figure 7: Illustration of the calculating correlation between Kylie and make-up

We can do exactly the same calculation for each and every one of the teenagers and sum them all up. This is what is meant by:

$$\Sigma_i(M_{i,x} - \overline{M_x})(M_{i,y} - \overline{M_y})$$

The Σ_i indicates that we should sum across all of the twelve teenagers. When we sum up all of the teenagers' attitudes to make-up, multiplied by their attitude to Kylie, we get:

$$2 - 8 + 20 - 16 + 10 + 8 + 6 + 2 + 20 + 0 + 9 + 16 = 69$$

Most of the numbers are positive, showing that the kids feel the same way about Kylie and make-up. Madison and Jacob are among those who contribute positive numbers, 2 and 20 respectively. The exceptions are Tyler, who doesn't like Kylie, and Ryan, who doesn't like make-up but does like Kylie Jenner. It is these two who are responsible for the −8 and the −16.

Mathematicians don't like big numbers like 69. We prefer our numbers to be small, preferably near to 0 or 1, so that we can compare them to each other. We achieve this through the denominator (the bottom part of the fraction) in Equation 7. I'm not going to go through the calculation in detail, but if we put in the numbers we find that:

$$r_{\text{make-up,Kylie}} = \frac{69}{\sqrt{120 \cdot 152}} = 0.51$$

This gives us a single number, 0.51, which measures the correlation between make-up and Kylie. A value of one would indicate a perfect correlation between the two types of posts, while zero would indicate no relationship at all. So the actual value of 0.51 suggests a medium-sized correlation between liking make-up and liking Kylie Jenner.

I realize I have been doing a fair bit of calculating now, but we still only have one of the fifteen important numbers we need to know about our teenagers' snaps! We don't just want to know the correlation between make-up and Kylie, but we would like to know the correlation between all the different categories: food, make-up, Kylie, PewDiePie, *Fortnite* and Drake. Luckily, now we know how to calculate one correlation using Equation 7, it is just a case of putting pairs of categories into the equation one after another. And that's exactly what I'll now do. This gives us what is known as the correlation matrix, which we denote R.

	Food	Make-up	Kylie	PewDiePie	*Fortnite*	Drake	
	1.00	0.24	0.23	−0.61	−0.10	−0.11	Food
	0.24	1.00	0.51	−0.63	−0.74	−0.26	Make-up
$R =$	0.23	0.51	1.00	−0.17	−0.17	−0.69	Kylie
	−0.61	−0.63	−0.17	1.00	0.71	−0.08	PewDiePie
	−0.10	−0.74	−0.17	0.71	1.00	0.06	*Fortnite*
	−0.11	−0.26	−0.69	−0.08	0.06	1.00	Drake

If you look in the row labelled 'Kylie' and the column labelled 'Make-up', you will see 0.51, the correlation we just calculated. The other rows and columns are calculated in exactly the same way, with different pairs of categories. For example, *Fortnite* and PewDiePie have a correlation of 0.71. Other correlations, like *Fortnite* and make-up, with −0.74, are negatively correlated, which means that gamers are typically not so interested in make-up.

The correlation matrix groups people into stereotypes. When I asked you earlier to picture the teenagers in your head and not to be scared of being judgemental, I was essentially asking you to build your own correlation matrix. The Kylie/Make-up correlation puts people like Madison, Alyssa, Ashley and Kayla in one stereotype, the PewDiePie/*Fortnite* correlation puts Jacob, Ryan, Morgan and Lauren in another stereotype. Others, like Tyler and Matt, don't quite fit into any simple category.

In May 2019, I spoke to Doug Cohen, data scientist at Snapchat, about the information they store in correlation matrices about their users.

'Well, that would be pretty much every activity you perform on Snapchat,' he answered. 'We look at how often our users chat with friends, the number of streaks they have, the filters they use, how long they spend looking at maps, the amount of group chats they are involved in, then how long they spend looking at content or when they look at their friends' stories. And we look to see how these activities correlate with each other.'

The data is anonymized, so Doug doesn't know exactly what you as an individual are up to. But these correlations allow Snapchat to categorize its users as everything from the 'selfie obsessed' and 'documentary makers' to the 'make-up divas' and the 'filter queens', to use some of the company's own marketing terminology.[2]

Once they know what gets a user engaged, they can give them more of the same. Listening to Doug talk about his work on increasing engagement, I couldn't help commenting. 'Wait a minute here,' I said, 'as a parent, I'm trying to get my kids to use their phones less and you are working at the other end to maximize their engagement!'

Doug defended himself with a small dig at his rivals. 'We don't just try to maximize time spent on the app, like Facebook have done traditionally,' he said, 'but we do look at participation rate, how often our users return. We help our users touch base with friends.'

Snapchat don't necessarily want my kids to spend all of their time on their app, but they do want them to keep coming back for more. And I can tell you from personal experience that it works.

*

Most of us want to be respected as individuals rather than being portrayed as stereotypes. Equation 7 totally ignores our wishes. It reduces us to correlations between the things we like.

The mathematicians working at Facebook realized the power of correlations at an early stage of the platform's development. Every time you like a page or comment on a subject, your activity provides data to Facebook on you as an individual. The way Facebook has used this data has evolved over time. In 2017, when I first started to observe how the analysts were monitoring us, the categories were quite fun: 'Britpop', 'royal weddings', 'tugboats', 'neck' and 'upper middle class' were some of the boxes they put us into.

These categories left many Facebook users feeling uncomfortable and, more importantly for the company's bottom line, they weren't particularly useful for advertisers. By 2019, Facebook had revised its labelling to be more product-specific. Dating, parenting, architecture, war veterans, environmentalism are a few of several hundreds of categories used by the company to describe its users.

One reaction to being stereotyped in this way is to say that it is wrong; to shout, 'I am not a data point, I am a real person, I am an individual.' I'm sorry to break it to you, but you are not quite as unique as you might like to think you are. The way you browse has found you out. There is someone else out there with the same combination of interests as you, with the same favourite photo filter as you have, who takes as many selfies as you do, follows the same celebrities as you do and who clicks on the same ads as you do. In fact, there isn't just one person; there are lots of them, all clumped together by Facebook, Snapchat and all the other apps you use.

There is no point getting upset or angry about the fact that you are a data point in the matrix. You should embrace it. In order to see why, we need to think about grouping people from a different perspective, a rather less appealing way of putting people into categories.

Imagine that the matrix M contains Madison, Tyler and the other kids' genes instead of their social media interests. Modern geneticists do indeed view us as data points: a matrix of 1s and 0s, indicating whether or not we possess certain genes. This correlation-matrix view of people saves lives. It allows scientists to identify the cause of diseases, to find personalized medicine tailored to your DNA and to better understand the development of different forms of cancer.

It also allows us to answer questions about our ancestral origins. Noah Rosenberg, a researcher at Stanford University, and his colleagues constructed a matrix of 4,199 different genes and 1,056 people from all over the world. Each of these genes differed between at least two of the people in the study. This is an important point because all humans have lots of genes in common (these are the genes that make us human). Rosenberg was specifically looking for differences between humans and how our place of origin contributes to differences between us. How do Africans differ from Europeans? And how do people from various parts of Europe differ from each other? Are the differences in our genes explained by what we commonly refer to as race?

To help answer this question, Noah first used Equation 7 to calculate correlations between people in terms of the genes they shared.[3] Then he used a model called ANOVA (short for analysis of variance) to look at whether our geographical place of origin explains these correlations. There isn't a yes or no answer to this question: ANOVA gives a percentage answer between 0% and 100%. Would you like to guess how much of our genetic make-up is explained by our ancestral origins? 98%? 50%? 30%? 80%?

The answer is around 5% to 7%. No more than that. Other studies have confirmed Noah's findings. While some genes produce very noticeable differences between races, the most striking example being the genes that regulate melatonin production and skin colour, the concept of race is highly misleading when it comes to categorizing us. The geographical origins of our ancestors simply don't explain the differences between us.

It might be slightly patronizing of me to explain the naivety of race biology in 2020, but unfortunately some people do believe that certain races are, for example, inherently inferior in intelligence. These people are racists and wrong. There are others, the 'I'm not a racist, but . . .' types, who believe that accepting equality between races is somehow imposed on us by teachers or society. The retired professor I corresponded with, who wrote for *Quillette* (see Chapter 3), is one of these people. They believe that we suppress discussions about the differences between races in order to be politically correct.

In fact, the place our ancestors came from accounts for only a tiny proportion of variation in our genes. Furthermore, genes do not fully determine who we are as individuals. Our values and behaviour are shaped by our experiences and the people we meet. Who we are has little or nothing to do with biological race or our ancestral origins.

The under-twenties – like my imaginary teenagers Jacob, Alyssa, Madison and Ryan – make up the new Generation Z, following on from the Millennials. Being seen as individuals is extremely important to this new generation. They certainly don't want to be viewed in terms of their gender or their sexuality. One survey of 300 individuals from Generation Z in the USA found that only 48% identified as completely heterosexual, with one third of those asked preferring to put themselves on a scale of bisexuality.[4] Over three-quarters of them agreed that 'Gender doesn't define a person as much as it used to'. My age group, Generation X, can be heard expressing scepticism about Generation Z's 'refusal' to see gender differences. Again, there is a conception that the Zs are trying to be politically correct and in the process denying basic biological facts.

There is another way of seeing this generational change. The Z generation have a lot more data than people of my age had when we were young. While Xers grew up with a limited number of stereotypes provided by TV shows and our limited personal experience, the Zers are bombarded with imagery of diversity and individuality. Generation Z see this individuality as more important than maintaining gender stereotypes.

The success of Facebook's advertising categories, based on correlations in our interests, suggests that Generation Z's view of the world is statistically correct. Douglas Cohen, who worked in advertising at Facebook before moving to Snapchat, told me that his former employer's advertisers bid against each other to hyper-target their adverts on the small interest groups identified in the company's correlation matrix. The price to reach a targeted audience can double or triple as advertisers compete for the rights to talk directly to DIY enthusiasts, action-film fans, surfers, online poker players and many other interest groups. To the advertisers, individual identity is worth a lot of money.

Properly categorizing people according to the things they really like and the activities they enjoy can be extremely effective and fair. Correlations can help us find groups with common interests and aims, just as scientists use correlations between genes in order to find the causes of diseases.

*

The Houses of Parliament can be an intimidating place for a young data scientist. 'It isn't long since the signs in Westminster referred to members of the public as "strangers",' Nicole Nisbett told me when I met up with her at the University of Leeds. 'It's changing, and the staff at Westminster are actively reaching out to the public and researchers, but these signs reveal an historical wariness of outsiders.'

Nicole is two years into a PhD project, based half the time in Leeds and half the time at the House of Commons, which means she now has an 'access most areas' pass to the House. She is on a mission: to improve how Members of Parliament (MPs) and their permanent staff listen and engage with the outside world. Before Nicole started her project, many of the staff, who deal with the day-to-day running of government, felt that all the comments the public made on Facebook or even on their own discussion forums were too overwhelming to engage with. 'There was also a feeling that they already knew what people were going to tell them,' Nicole said, 'and sifting through all the negative comments and abuse made it an arduous task.'

Nicole's background in data science has given her a different perspective. She understood how the number of comments on Twitter and Facebook could be overwhelming for any one individual, but she also knew how to find correlations. She showed me a map she had created that summarized a debate on banning animal fur products. She put all of the words used in the debate into a matrix and looked to see how their usage was correlated. Words that were used together were linked together. 'Fur' was found in a cluster with 'selling', 'trade', 'industry', which further connected to 'barbaric' and 'cruel'. Another cluster linked the words 'suffer', 'killing' and 'beautiful'. A third cluster linked 'welfare', 'laws' and 'standards'. Each of these clusters summed up a strand of the argument.

In one area of Nicole's map, two words appeared side by side: one was 'electrocution', the other, 'anal'. A thick line connected them. I stared at the words, trying to work out what Nicole wanted me to infer. 'At first we thought these words were being used by trolls,' Nicole told me. In any debate there are always people on one side of the argument trying to wind up those on the other side, often with abusive language. But abusive discussions tend to have less correlation between the words used – abuse is, quite literally, random – whereas these two words were used repeatedly by a wide range of different people. When Nicole looked at the sentences containing them she found a very well-informed group of people discussing a genuine issue: farmed foxes and racoons were being killed by inserting electrical rods into their bodies and applying a very high voltage. This added a new dimension to the discussion by parliamentary staff, one that they would never have noticed without Nicole's work.

'I avoid making assumptions about what the public will write; my job is to condense thousands of opinions so that Parliament can react more rapidly to the debate,' Nicole told me. Different views feature in her analysis not because it is 'politically correct' to hear all sides of the argument, but because it is statistically correct to highlight important opinions. Minority views are given a voice because they genuinely contribute to the conversation. Correlations give an impartial representation of all sides of the argument without our having to take a political position on which views should and shouldn't be heard.

'It is baby steps. We can't solve everything with statistics,' Nicole told me. Then she laughed, 'No amount of data analysis can help with Brexit!'

*

Social scientists go to great lengths to find statistically correct explanations on data. I first got to know Bi Puranen, researcher at the Institute for Future Studies in Stockholm, when we travelled to St Petersburg together to speak at a conference on political change. The researchers at the institute we visited were financed directly from money allocated by Dmitry Medvedev, Russian president during Putin's time as prime minister. But the sentiments of the young PhD students employed there were firmly anti-establishment. They were desperate for democratic change and spoke to us passionately about how their views were suppressed. I witnessed at first hand how Bi carefully navigated the conflicts, sympathizing with the students while accepting the reality of conducting a research project within Putin's Russia.

For Bi it was vital that, irrespective of their political views, the Russian researchers she worked with carried out the World Values

Survey in exactly the same way it was carried out in every country that participated (almost one hundred altogether). By asking people from all over the world the same set of questions, many of which were about sensitive subjects such as democracy, homosexuality, immigration and religion, Bi and her colleagues wanted to understand how the values of the planet's citizens vary between countries.[5] This was a point that even the most politically motivated researcher quickly understood: data should be collected in as neutral a way as possible.

There are 282 questions in the survey, so correlations provided a useful way of summarizing similarities and differences between the answers given. Two of Bi's colleagues, Ronald Inglehart and Christian Welzel, found that people who emphasize family values, national pride and religion tend to have moral objections to divorce, abortion, euthanasia and suicide. Correlations in the answers to these questions allowed Inglehart and Welzel to classify citizens of different countries on a traditional–secular scale.[6] Countries like Morocco, Pakistan and Nigeria tended to be traditional, while Japan, Sweden and Bulgaria tended to be secular. This result by no means implied that everyone in a country has the same views, but it gave a statistically correct summary of the views prevalent in each nation.

Chris Welzel went on to find another way in which answers to the questions were correlated. Those people who are concerned about freedom of speech also tend to value imagination, independence and gender equality in education, and had a tolerance for homosexuality. Answers to these questions, which Welzel called emancipative values, were positively correlated. Britain, the United States and Sweden are among the countries that have high emancipative values.

A really important point here is that the first axis, traditional/secular, was *not* correlated with the second axis, emancipation. At the start of this millennium, Russians and Bulgarians, for example, had high secular values, but did not value emancipation. In the USA, freedom and emancipation are important for almost everyone, but the country remains traditional, in the sense that religious and familial values are prioritized by many citizens. The Scandinavian countries are the extreme examples of both secular and emancipative values, while Zimbabwe, Pakistan and Morocco are at the opposite extreme, valuing both tradition and obedience of authority.

The separation of values on two independent axes gave Bi Puranen an idea. She wanted to know how immigrants' values changed when they arrived in Sweden. In 2015, 150,000 immigrants sought asylum in Sweden, mainly from Syria, Iraq and Afghanistan. This number represented roughly 1.5% of Sweden's total population, and they all arrived within one year from three countries with completely different sets of cultural values to those in their new home.

When natives of West European countries look at these immigrants, they often notice things that relate to their traditional values: for example, the hijab or a newly built mosque. These observations lead some to conclude that Muslims, in particular, are somehow failing to adapt their values to their new homeland. External appearances might show how the immigrants are trying to preserve their own traditions, but the only statistically correct way to really understand their internal values is to talk to them and ask them what they think. This is exactly what Bi and her colleagues did. They surveyed 6,501 people who had arrived in Sweden over the last ten years and asked them about their values.

The results were striking. Many of those immigrants shared the typical European's desire for gender equality and tolerance for homo-sexuality, while not adopting Sweden's extreme secularity. They maintained their traditional values – those that are visible to outsiders, relating to the importance of the family and religion. In fact, a typical Iraqi or Somali family living in Stockholm has very similar values to an all-American family living in Texas, USA.

Muslims are not the only minority group that isn't understood in the statistically correct way mentioned above. I often hear anti-abortion and homophobic views being lumped together when people are talking about Christian Americans for example. Michele Dillon, Professor of Sociology at the University of New Hampshire, has shown that some anti-abortion religious groups were for gay marriage, while other religious groups had the opposite set of opinions.[7] In general, abortion and gay rights are regarded as separate issues within religious groups.

*

As more parts of our lives have moved online, the data available about us has grown too: who we interact with on Facebook, what we like, where we go, what we buy, the list goes on and on. Every social interaction, search query and consumer decision is stored within Facebook, Google and Amazon. This is the world of big data. We are no longer defined by our age, gender or town of birth, but by millions of data points measuring our every move and thought.

TEN moved quickly to take up the big data challenge. Its members matrixed the world's population. They connected people based on their interests. They thought they had shown that racism and sexism were things of the past. They measured how society was evolving to become a more tolerant place – a world that was fair and respected individuals for who they really were. TEN was being statistically correct.

Much of the new order was financed by advertisements, tailored to the individual. Advertisers entered into bidding wars for the right to show their products to small focused groups of Facebook users. More data scientists and statisticians were recruited to help deliver information with greater precision. A new field of micro-targeted advertising was born. Potential customers were profiled and fed just the right information at the right time to maximize their interest.

The members of TEN had won again, adding advertising and marketing to the list of problems they had solved. This time it even seemed that they had some form of morality on their side. But there was a problem. It wasn't just the members of TEN who were staring at the numbers in the matrix. And not all of the people looking at its correlations properly understood the patterns that they saw . . .

*

Anja Lambrecht's research is about using big data correctly. As Professor of Marketing at the London Business School, she has studied how data is used in everything from brand clothing to sports websites. She explained to me in an email that, while there were obvious benefits to using big data sets in advertising, it was also important to consider the limitations. 'Data without the skills to extract the appropriate insights are not very helpful,' she told me.

In one of her scientific articles, from which the following example is adapted, Lambrecht, together with her colleague Catherine Tucker, explains the problems using a scenario from online shopping.[8] Imagine a toy retailer which finds that consumers who see more of their adverts online buy more of their toys. In doing so they have established a correlation between their advertisements and toy purchasing, and their 'big data' marketing department concludes that their advertising campaign works.

Now look at the adverts from a different angle. Consider Emma and Julie, who don't know each other but both have seven-year-old nieces. Independently, they see the company's toy adverts while browsing on the last Sunday before Christmas. Emma has a busy week ahead at work and doesn't have time to shop. Julie is on holiday and spends a good part of her down time browsing for Christmas presents. After seeing a retailer's advert three or four times for the *Connect 4 Classic Game of Counter Strategy*, Julie clicks on it and decides to buy. Emma ends up going to a store on the afternoon of 23 December and grabs a Lego model camper van and buys it.

Julie has seen the advert a lot more times than Emma, but does that mean the advert is effective? No. We have no idea what Emma would have done if she had had time to watch the adverts. Those 'big data' marketers, who might have concluded that their campaign works, have confused correlation with causation. We don't know if the advert caused Julie to buy *Connect 4*, so we can't conclude that it is effective.

Separating causation and correlation is difficult. The correlation matrix I created above for Madison, Ryan and their friends is based on a very small number of observations, so we can't draw any general conclusions from it alone (remember the confidence equation?). But imagine that data for the same matrix were obtained for the views of large numbers of Snapchat users and we did find that PewDiePie is correlated with *Fortnite*. Can we then draw the conclusion that helping PewDiePie gain more subscribers would increase the number of *Fortnite* players? No, we can't. To jump to this conclusion would again confuse correlation with causation. Kids don't play *Fortnite* *because* they watch the YouTuber PewDiePie. A campaign to increase PewDiePie's subscriptions will (if it works) do exactly that, it will increase the amount of time kids spend watching PewDiePie. It won't cause them to play a game of *Fortnite* after they have watched his videos.

What if *Fortnite* buys advertising space on PewDiePie's channel? This might work: maybe some *Fortnite* players have drifted back to *Minecraft* and PewDiePie can lure them back again. But it could also very well fail. It could be the case that the interest in *Fortnite* among PewDiePie viewers is already at saturation point. Maybe the campaign should instead focus on getting Kylie Jenner to start playing the game!

A bit of thought allows us to see all sorts of problems with drawing conclusions from the PewDiePie/*Fortnite* correlation in our data. But when the big data revolution started, many of these problems were ignored. Companies were told that their data were extremely valuable to them, because now they knew everything about their customers. They didn't.

*

Cambridge Analytica was a clear example of a company that had failed to account for causation.

The Senate Committee listened carefully as I talked to them over Skype. 'Cambridge Analytica collected a lot of data on Facebook users, in particular the products and sites they clicked the "Like" button for. Their aim was to use this to target Facebook users' personalities. They wanted to target neurotic individuals with messages about protecting their family with a gun, and traditional individuals with messages about handing guns down from father to son. Each advert would be tailor-made to the voter.'

I was aware that the group I was talking to now – Republican committee members – could well be imagining the advantages of having such a tool in the next election campaign. So I got to the point quickly. 'But it couldn't work, for several reasons,' I said. 'First it just isn't possible to reliably work out people's personalities from their "Likes". Their targeting would have got the personality of people wrong only slightly less often than it got it right. Secondly, the type of neuroticism that can be found out about Facebook users – those that like

Nirvana and an Emo lifestyle – is different from the neuroticism associated with protecting your family with weapons.'

I went through the problems arising from confusing correlation and causation. When Cambridge Analytica created their algorithm, the election hadn't happened yet, so how could they test whether or not their adverts were working?

I went on to tell them about the ineffectiveness of fake news in influencing voters, another thing I had been researching for my previous book, *Outnumbered*.[9] I also told them that, contrary to the echo-chamber theory, Democrat and Republican voters in the 2016 election had got to hear all sides of the story. My view contradicted much of the consensus in the liberal media at the time, which saw Trump's win as a victory for online voter manipulation. His voters were accused of being naive and the victims of brainwashing. Cambridge Analytica had come to represent the ease with which public opinion could be swayed by social media. I didn't agree with the consensus.

The person coordinating the call said, 'I'm just going to put you on mute while we discuss what we've heard.'

It took them about thirty seconds to reach a decision. 'We would like to fly you to Washington to testify to the Senate Committee. Would you be able to come?'

I didn't reply straight away. I mumbled something about having a holiday booked and told them I'd have to think about it.

At that point I really wasn't sure whether I should go or not. But after sleeping on it, I felt increasingly sure that I shouldn't. I realized that they didn't want me to go to the USA to explain causation and correlation to the senators. They wanted me to go because I was going to say that Cambridge Analytica and fake news didn't get Trump elected. They wanted to hear those of my conclusions that fitted in with their narrative, rather than understand the models I was using. So I didn't go.

I did go to the USA that summer, though. I visited New York City and met up with Alex Kogan, just after he had testified at the Senate hearing. Alex, a researcher at the University of Cambridge, was seen as one of the bad guys in the Cambridge Analytica story. He had downloaded the data for 50 million Facebook users and then sold it to Cambridge Analytica. Not a particularly smart move and something he now regretted doing.

Alex and I had come into contact when I started to look at the accuracy of the Cambridge Analytica methods. I liked talking to him. When it came to doing business, maybe he wasn't keeping the best company, but he had a deep understanding of how data can and can't be used. He had really tried to create what Chris Wylie had called a 'psychological warfare' tool to pinpoint voters, but had concluded that the weapon simply couldn't be created. The data wasn't good enough.

Working on the inside, he had reached the same conclusion about Cambridge Analytica as me. 'That shit doesn't work,' he'd told me. At the Senate hearing, he had told the senators the same thing in more polite terms.

*

The basic 'problem' with the Cambridge Analytica algorithm was that it didn't work.

At the start of the 'big data' era many so-called experts had suggested that correlation matrices could lead directly to a better understanding of their users and customers. It isn't quite as straightforward as that. Algorithms based on correlations in data were used not only for political advertising, but also for giving advice on prison sentencing, for evaluating school teachers' performances and for finding terrorists. The title of Cathy O'Neil's book *Weapons of Math Destruction* nicely captures the problems that resulted.[10] Like nuclear bombs, algorithms are indiscriminate. While the term 'targeted advertising' suggests a tight control over who is shown adverts, in fact these methods have only a very limited ability to classify people correctly.

For online advertising this isn't too much of a problem. A *Fortnite* player's life isn't going to be ruined if he is shown an advert for make-up. But being labelled a criminal, a bad teacher or a terrorist by an algorithm is another matter. Careers and lives can be changed. Algorithms based on correlations were presented as objective, because they came from data. In fact, as I found when writing my last book,

Outnumbered, quite a number of algorithms made almost as many mistakes as they made accurate predictions.

I found a host of other problems that can arise when we create algorithms using correlation matrices. For instance, the way Google represents words inside its search engine and translation services is built on correlations between word usage.[11] Wikipedia and databases of news articles are also used to identify when certain groups of words are used together.[12] When I looked at how these language algorithms saw my name, David, compared to Susan, the most popular name for a woman my age in the UK, it drew some very unflattering conclusions. Where I was 'intelligent', 'brainy' and 'smart' as 'David', the algorithm labelled Susan as 'resourceful', 'prissy' and 'sexy'. The underlying reason for this problem is that the algorithms are built on correlations in the historic texts we have written, which are filled with stereotypes.

The algorithms used for big data found correlations, but they didn't understand the reasons for these correlations. As a result, they made massive mistakes.

*

The consequences of overselling 'big data' were complicated, but the causes were simple. Remember how we divided the world up into data, model and non-sense? What happened was that companies and the public were told about the data without any proper discussion of the models. When the models are missing, the non-sense takes over. Alexander Nix and Chris Wylie were talking nonsense about personality targeting and psychological warfare tools. Companies predicting teacher performance and creating sentencing software were talking nonsense about the effectiveness of their products. Facebook was reinforcing false stereotypes with its ethnic-affinity-targeted adverts.[13]

Anja Lambrecht has a solution. She solves the causation problem by introducing a model – by creating a story – like her story about Emma and Julie and their ways of shopping. By taking the viewpoint of customers, instead of just looking at the data collected, we can assess the success of an advertising campaign. Although she doesn't describe it this way, Lambrecht is breaking the problem down into model and data, a strategy we have used throughout this book. Data by itself tells us very little, but when put together with a model we can gain insights.

This basic modelling approach to determining causation is known as A/B testing. I have already outlined this method in Chapter 1, and now we can put it into practice. A company should try two different adverts on its customers: (A) the original advert whose effectiveness they wish to test, and (B) a control advert, for a charity, for example, which contains no reference to the toy company. If the company sells just as many products to customers who see the charity advert as to those who see the original, then they know that their adverts have no effect.

Anja Lambrecht's research provides a wealth of examples of the way we should approach causation. In one study she investigated the idea commonly held in advertising that getting the attention of early-stage influencers on social media should help products go viral.[14] If an advertiser targets people who are quick to adopt new trends, then the advert should have a greater effect. Surely that makes sense, doesn't it?

To test the idea properly, Lambrecht and her colleagues compared an (A) group of users who, early on, shared hashtags with the latest trends, such as #RIPNelsonMandela and #Rewind2013, with a (B) group of users who were late to post on the same trends. Both A and B groups were shown a sponsored link to an advert and the researchers measured whether they clicked on or retweeted (shared) the advert.

The 'early-stage influencer' theory proved to be incorrect. The A group was less likely to share or click on an advert than the B group. These results were the same, irrespective of whether the advert was for a homeless charity or a fashion brand. Influencers are difficult to influence. They are influencers *because* they discriminate before they share online. Their independence and good judgement may well be part of the reason others follow them in the first place. An early adopter without good judgement is just a spammer, and no one wants to follow a spammer.

At Snapchat, Doug Cohen and his team A/B test everything. When I spoke to him, they were working on notifications, testing a complex variety of schemes to find out which of them encourages users to open up the app. But he was cautious about how well they understand their users: 'You are a different person in the morning than at the end of the day. So, we can put you in a broad category, but you actually change through the week, through the month and through the year, as you get older.' He also emphasized that people don't just want to see the same things all of the time: 'We might classify someone as interested in sport, but that doesn't mean they want to just see a load of masculine stuff.' Users get irritated if they think an algorithm is putting them into a certain type of category.

The advertising equation tells us that a certain amount of stereotyping is an unavoidable consequence of organizing large amounts of data. So, don't get upset about being part of a correlation matrix. It is a truthful representation of who you are. Look for correlations in your friends' interests and build connections. When correlations are genuine (and not built on racial or gender stereotypes), they make it easier to find points of commonality. If there are exceptions to the rule, be accepting and readjust your model. Look for patterns in your conversations, the way Nicole Nisbett did in political discussions, and use it to simplify debates. Look carefully for small clusters of new and novel points of view and pay special attention to these. Don't confuse correlation and causation, though. When you invite friends over for dinner, A/B test the menu. Don't just keep making pizza because they said it was nice last time. Build a statistically correct model of your world.

*

On the subway through Manhattan, after talking to Alex Kogan in New York, I realized something about myself. I no longer knew my own political position. I have always been left-wing and, since I read Germaine Greer when I was nineteen, I have been a feminist. I have always been what is now called 'woke'. Or, at least, as woke as a white boy who grew up in a small working-class town in Scotland in the 1980s can be. Being left-wing, disapproving of Donald Trump and blaming the manipulation of people through social media for his election were seen by many as part of a package. The media then portrayed them as correlated properties of a left-wing view of the world.

But maths had forced me to accept a different model. I was forced to grant Trump his victory, because the model said that it was a deserved victory. I didn't approve of his political views, but neither did I approve of the way his voters were portrayed as stupid and easily manipulated. They weren't.

Instead of looking for the real reasons for the rise of nationalistic sentiment that had brought Donald Trump to power, that had caused Brexit, that was responsible for the Five Star Movement in Italy, Viktor Orbán in Hungary and Jair Bolsonaro in Brazil, everyone seemed to be rushing to find a James Bond-type villain, an evil individual who had poisoned the political waters. Their Dr No came in the form of Alexander Nix and his company Cambridge Analytica. Somehow this man, with only a basic understanding of models and data, was meant to have manipulated the whole of modern democracy.

The biggest threat to any secret society is the threat of being uncovered. Cambridge Analytica had been uncovered and its threat revealed. It was a threat equivalent to that of a mediocre advertising company paid, at most, $1 million by Trump in an election in which $2.4 billion was spent. The effect was more or less proportional to the size of the investment: minuscule.

The James Bond villain conspiracy theory lacked judgement. There was very little in the data to give us any confidence that it was true. TEN, on the other hand, continued to operate unnoticed. Members run banks, financial institutions and bookmakers. They create our technology and control our social media. In each of these activities they take a little cut, 2 or 3 cents to the dollar in gambling; 1 cent to the dollar in online trading; and even less when they provide an advert in combination with the results of an Internet search. Over time the small cuts add up and the profits amass for TEN's members. In all parts of life, mathematicians are out-competing those who don't know the equations.

Sitting on the subway travelling home, I thought about the people who had been begging me for football tips and about the ways in which online betting companies offered lonely men the chance to chat to women while they were gambling their money away. I thought about the prism of Instagram through which we saw a lifestyle based around consumerism and celebrity. I thought about the adverts for high-interest mobile phone loans that target the poorest sections of society.

The underlying tensions in our society, specifically those between rich and poor, were being ignored by people who explained away the outcome of the US presidential election and Brexit as attributable to fake news and Cambridge Analytica's use of stolen Facebook data. People like me, mathematicians and academics, have also played a big part in making society less fair. Members of TEN are taking an edge from the poor and making themselves rich. The irony is that one conspiracy theory really is true. The Illuminati does exist, in the form of TEN. It is so deeply hidden that even the co-conspirators can't see it.

8

The Reward Equation

$$Q_{t+1} = (1 - \alpha)Q_t + \alpha R_t$$

I spent fifteen years of my early professional life investigating how animals look for and collect rewards. It wasn't something I had decided I had to do – it was an activity I somehow fell into.

Two biologist friends, Eamon and Stephen, asked me if I'd like a day trip to Portland Bill, a narrow peninsula off the southern coast of England. They needed more ants. Eamon showed me how to carefully prise open a small rock crevice where he thought they could be hiding. He seemed to get it right every time, finding ants under his chosen stones. He quickly sucked them up through an improvised vacuum cleaner into a test tube that was to be taken back to the lab.

It took time, but eventually I got some, standing some distance away from my colleagues, the lighthouse high above me. It was a rewarding feeling, sucking an orange plastic tube to vacuum up an ant colony. We spent five years studying how these ants choose a new nest. I made the models; they collected the data.

The walks I took with Madeleine, then a post-doctoral researcher in biology in Sheffield, on the Yorkshire moors were for a similar purpose: this was where honey bees would fly, up to 8 miles from their hive, to collect pollen from the rich heather. Up there, our heads cleared of the thick office air, I'd try to relate my equations to Madeleine's descriptions of how bees and ants communicate about food. We worked together for over ten years, looking at how different species of social insects decide which food sources to exploit.

Many discussions took place in less glamorous places. Dora, then a PhD student at Oxford and the first friend I made when I moved there, told me about her pigeons as we sat on a cold step next to a kebab van. A few days later we were poring over GPS tracks of her birds in the Jericho Café. A year later we were putting the finishing touches to an article about how pairs of birds compromised on their routes when flying home.

Ashley carefully constructed Y-mazes for sticklebacks. I met him in the pub with Iain and we talked about how we could model their decisions in groups. Together, we looked at how they avoided predators and followed each other to food.

My journey then moved out of England and further afield. Big-headed ants in Australia with Audrey. Argentine ants with Chris and Tanya. Leaf-cutter ants in Cuba with Ernesto. House sparrows in the south of France with Michael. Locusts from the Sahara with Jerome and Iain. Slime moulds in Japan with Toshi (and with Audrey and Tanya too). Cicadas in Sydney with Teddy.

All of my colleagues from that time are now professors in universities around the world. But that was never our sole aim. We were (and still are) people who talked to each other, learnt from each other and solved problems together. We collected small rewards by answering questions and slowly we came to a better understanding of the natural world. By the end of fifteen years, I knew pretty much everything there was to know about the way animals make decisions in groups. It wasn't completely clear in my mind at that time, but looking back on it now I realize that there was one equation behind almost everything I accomplished then.

*

Animals need only two things to survive: food and shelter. And they need one more thing to reproduce: a mate.

Underlying all three of these requirements for life, there is an even more fundamental thing that animals need to obtain: information. Animals gather information about food, shelter and sex from their own experiences and from the experiences of others. They then use this information to survive and reproduce.

One of my favourite examples is ants. Many ant species leave pheromone, a chemical marker to show their nest-mates where they have been. When they find a sugary food lying on the ground, they deposit their pheromone. The other ants search out this pheromone and follow it to the food. The result is a feedback mechanism, whereby more ants leave pheromone and find the food more quickly.

Humans also need food and shelter to survive and a mate to reproduce. In our evolutionary past, we spent a lot of our time searching out the information that would allow us to obtain and retain these three essentials. In modern society the search has changed form. For a fair proportion of the world's population, the search for essentials is over, but the search for information about food, housing and sex has continued and expanded: it now takes the form of watching cooking shows and *Love Island*; reading celebrity gossip; browsing houses for sale and property prices. We post pictures of our partners, our dinner, our children and our homes. We show each other where to go and what to do. Like ants, we do everything possible to share what we have found out and follow the advice that has been shared.

I am a bit embarrassed to admit the extent of my daily information search. I go to Twitter to check my notifications; I open my mail for new messages; I read the political news; and then I start to click through the sports news. I go into the online publishing platform Medium to see if anyone has liked my stories and check whether there are any interesting comments.

The mathematical way of interpreting my behaviour brings us back to the slot machines we looked at in Chapter 3. Each of the apps on my phone is like pulling a handle and seeing if I get a reward. I pull the Twitter handle: seven re-tweets! I pull the email handle: a message inviting me to give a talk. Yay, I'm popular. I pull the news and sports handle: another Brexit intrigue or a transfer rumour. I go into Medium – but no one has liked my posts. Oh dear, that handle doesn't work so well.

I'll now put my app slot-machine life into an equation. Imagine I open Twitter once every hour. This is probably an underestimate, but we need to start our model with a simple assumption.

I denote the reward I get at hour t as R_t. Again, to keep it simple, we say that $R_t = 1$ if someone has re-tweeted my post and $R_t = 0$ if no one has given my post a re-tweet. We can think of the rewards throughout a working day from 9 a. m. to 5 p. m. as a sequence of 1s and 0s. For example, they might look like this:

$$R_9 = 0, R_{10} = 1, R_{11} = 1, R_{12} = 0, R_{13} = 0, R_{14} = 1, R_{15} = 0, R_{16} = 1, R_{17} = 1$$

The rewards model the re-tweets of the external world.

Now, we need to consider my internal state. By going into the app, I am improving my estimate of the quality of Twitter, of its ability to give me that momentary feeling of self-confirmation that only a 're-tweet' or a 'like' can provide. Here we can use the reward equation:

$$Q_{t+1} = (1 - \alpha) Q_t + \alpha R_t$$

(Equation 8)

In addition to time t and reward R_t, this equation has two additional symbols: Q_t represents my estimate of the quality of the reward and α determines how quickly I lose that confidence in the absence of a reward. These letters require some additional explanation, so here we go.

If I wrote $Q_{t+1} = Q_t + 1$ then it would indicate that I am increasing Q_t by 1. This idea is used in computer programming inside a 'for loop': we increment Q_t by 1 each time we go through the loop. The same idea applies in the reward equation. In this case, instead of adding 1, we update Q_t by combining two different components. The first component, $(1 - \alpha)Q_t$, reduces our estimate of the quality of the reward. For example, if we set $\alpha = 0.1$ then at every time-step our estimate will decay by $1 - 0.1 = 90\%$ of its previous level. This is the same equation we use to describe how, for example, a car depreciates in value each year or, and this is going to be important later on, how pheromones and other chemicals evaporate. The second component, αR_t, acts to increase our estimate of the value of a reward. If the reward is 1, then we add α to Q_{t+1}.

Putting these two components together we can see how the equation as a whole works. Imagine that I start work at 9 in the morning with an estimate $Q_9 = 1$. I believe 100% that Twitter has the quality to provide me with a rewarding re-tweet. I open it up, but find, to my disappointment, that $R_9 = 0$. No reward. No re-tweets. So, I use Equation 8 to update my quality estimate to $Q_{10} = 0.9 \cdot 1 + 0.1 \cdot 0 = 0.9$. Now I am a bit less confident when I open up Twitter at 10 a. m., but this time I get what I am looking for: $R_{10} = 1$. A re-tweet! My estimate of the quality doesn't recover entirely, but it is edged back up a bit: $Q_{11} = 0.9 \cdot 0.9 + 0.1 \cdot 1 = 0.91$.

In 1951, mathematicians Herbert Robbins and Sutton Monro proved that Equation 8 always gives a correct estimate of the average value of the reward.[1] To understand their result, assume that the probability I get a reward (in terms of a re-tweet) on any particular hour is denoted by the symbol $\bar{R}$, and let's say that $\bar{R} = 0.6$ or 60%. Before I start my hourly checking of Twitter I have no idea what $\bar{R}$ is. My aim is to estimate its value from the sequence of rewards I get when I open the app. Think of the individual rewards that arrive as a series of 1s and 0s – like 011001011... . If this sequence continued indefinitely, the average frequency of 1s would be $\bar{R} = 60\%$.

Equation 8 quickly starts to reflect the rewards on offer: $R_{11} = 1$ so $Q_{12} = 0.919$, $R_{12} = 0$ so $Q_{13} = 0.827$, and so on until, at the end of the day, we get $Q_{17} = 0.724$.[2] Each observation takes me closer to estimating the true value of $\bar{R}$. For this reason, Q_t is often referred to as a tracking variable: it tracks the value of $\bar{R}$. Figure 8 illustrates this process.

Robbins and Monro showed that, in order to reliably estimate $\bar{R}$, we don't need to keep a note of the whole sequence of 1s and 0s. All we need to know in order to update our new estimate, Q_{t+1}, is our current estimate, Q_t, and the next reward in the sequence, R_t. As long as I have calculated everything correctly up to this point, I can forget about the past, and just store the tracking variable.

There are a few caveats. Robbins and Monro showed that we need to very slowly decrease α over time. Remember, α (Greek letter) is a parameter that controls how quickly we forget. Initially we are uncertain, so we should pay a lot of attention to the most recent values by setting α to a value close to 1. Then, over time, we should decrease α, so it becomes closer and closer to 0. It is this slow change that ensures our estimate is guaranteed to converge to the reward.

*

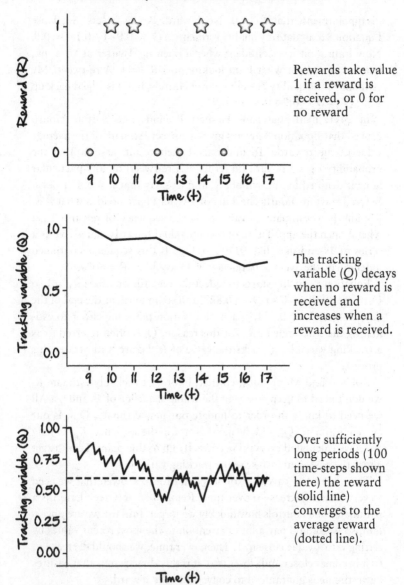

Rewards take value 1 if a reward is received, or 0 for no reward

The tracking variable (Q) decays when no reward is received and increases when a reward is received.

Over sufficiently long periods (100 time-steps shown here) the reward (solid line) converges to the average reward (dotted line).

Figure 8: How the tracking variable tracks rewards

Imagine you are lying on the couch, rewarding yourself by bingeing on TV. You start watching a Netflix series. The first episode is brilliant (as always), the second is OK and the third is slightly better. The question is how long should you keep watching before you give up? Your brain doesn't really care, but you do. You want to watch something good on your night off.

The solution is to apply the reward equation. For TV series, a good value for our decay in confidence is $\alpha = 0.5$, or one half. This is a very quick rate of forgetting the past, but TV is about being entertained now. A good show has to keep bringing new ideas.

Here is what you do. You give the first episode a mark out of 10, say 9. That is $Q_1 = 9$. If you are binge-watching, then just keep the number 9 in your head, and start the next episode. Now rank the new episode. Let's say it gets a 6. Now take $Q_2 = 9/2 + 6/2 = 7.5$. It's probably a good idea to round up each time, so the new ranking is 8. Watch the next episode. This time we'll say it got a 7. I take $Q_3 = 8/2 + 7/2 = 7.5$, which again I round up to 8.

Just keep going. The power of the method is that you don't need to remember how much you liked earlier episodes. Note the single number Q_t for the most recent episode in your head. Store a tracking variable Q_t, not only for how much you are enjoying a TV series, but also for how much you enjoy going to various types of social occasions, reading different authors or taking a Yoga class. This single number for each pastime allows you to understand – without going back to the time you got trapped talking to the boring mathematician at after-work drinks or when you pulled your sciatic nerve at yoga – the overall reward of different activities.

When should you stop watching? To answer this question, you need to set a personalized threshold. I use 7. If a series drops to a 7, then I stop. It's quite a brutal rule, because it means that if my series is on an 8 and I watch an episode with a 6 rating, then I have 8/2 + 6/2 = 7 and I am forced to stop watching. But I think it is fair. A good series should be ringing in the 8s, 9s and 10s on a regular basis. If it is hitting those highs then it will survive a 6, or even a 5. For example, if my current $Q_t = 10$ and I watch an episode that can only be described as a 5, then $Q_{t+1} = 10/2 + 5/2 = 7.5$, rounded up to 8 and I keep going. The next episode needs to be good, though. Based on this rule, I watched three and a half seasons of *Suits*, two seasons of *Big Little Lies*, one and a half seasons of *The Handmaid's Tale* and two episodes of the series *You*.

*

Most computer games use only one number – a score or a level – to track how well you are doing. The score is like Q_t in the reward equation: it monitors the rewards you have collected. You choose what you want to do next – which route you want to drive in *Mario Kart*, which adversary you want to track down and kill in *Fortnite*, which row you move on *2048*, which gym you raid in *Pokémon Go* – and your score is updated depending on the quality of your choices.

Your brain does something very similar. The chemical substance dopamine is often referred to as the brain's reward system, and we sometimes hear people say they were 'rewarded' with a dopamine boost. This image of rewards is not sufficiently nuanced, however. Over twenty years ago, the German neuroscientist Wolfram Schultz reviewed the experimental evidence on dopamine and concluded that 'Dopamine neurons are activated by rewarding events that are better than predicted, remain uninfluenced by events that are as good as predicted, and are depressed by events that are worse than predicted.'[3] So dopamine isn't the reward R_t, it is the tracking signal Q_t.[4] Dopamine is used by the brain for estimating rewards: it gives you your in-game score.

Games satisfy many of our basic psychological needs, such as demonstrating our competence at tasks and working together in groups.[5] One reason we keep on playing may be because of the way the games measure the completion of these tasks. Real life is messy. When we make decisions at work and home, the outcomes can be complicated and it is difficult to judge the rewards. In games, it is simple: if we do well, we pick up a reward; if we do poorly, we lose out. Games strip away the uncertainty and allow our dopamine system to get on with what it likes doing best: tracking rewards. The simplicity of scores, presented to us in terms of a single tracking variable as we play a game, mirrors the working of our biological reward systems.

The computer games industry has cracked the reward equation. One study, in which, each day after work, UK professionals either played *Block! Hexa Puzzle*, a *Tetris*-like puzzle game, or used *Headspace*, a mindfulness app, found that players of *Block!* recovered better from work-related stress. Emily Collins, a postdoctoral researcher at the University of Bath, who carried out the study, said afterwards, 'Mindfulness might be good for relaxation, but video games provide psychological detachment. You get these internal rewards and a real sense of control.'[6]

One game designer, Niantic, has used our desire to collect rewards to create games that get us out and about. Their most famous game, *Pokémon Go*, involves going out into the real world and 'collecting' small creatures, the Pokémons, using mobile phones. The game encourages its players to walk, both to find the Pokémon and to hatch Pokémon eggs, and to work in teams. If you have seen a group of people standing outside your local church or library, tapping madly on their phones, they are very likely to be Pokémon hunters, gathered there to 'take down' a Pokémon gym.

I'm now going to tell you something personal. It is about my wife. Lovisa Sumpter is a very successful woman. She is an associate professor in mathematics education at Stockholm University. She teaches the students who will one day teach in high schools; she organizes and speaks at massive international conferences; she supervises Masters and PhD students; she writes reports which shape educational policy and gives inspirational talks to teachers. She is also a qualified yoga instructor. I could write a whole book about how amazing my wife is, a big part of which would be about how she has put up with me for all this time and shaped our family life.

That isn't the bit that is personal. Everyone who has met Lovisa knows how amazing she is. It is hardly a secret – her brilliance is an established empirical fact. The bit that is personal is that Lovisa has achieved all of this while living with chronic pain since 2004. In 2018 she was diagnosed with fibromyalgia, a condition that is characterized by long-term pain throughout the body; it is primarily a problem with the nervous system. Lovisa's body keeps sending pain signals to her brain. Her pain-tracking system then sends warnings instead of rewards. Every small ache or twinge becomes amplified, making it difficult to sleep, to concentrate and to remain patient with those close to her. There is no known cure. It is for this reason that *Pokémon Go* became an important part of Lovisa's life: it provides a place where she can obtain some of the rewards that her body is denying her.

The game has allowed Lovisa to focus on something else while she is in pain and it also ensures that she walks a lot every day. Through the game Lovisa has found lots of new friends, who she 'takes down gyms' and 'raids' with. Many of these raiders have stressful jobs, for example, working as nurses and doctors at the hospital. There are also teachers and IT specialists, students and other young people. At least one couple got together through Lovisa's group. And there are quite a few people who might have been excluded in other social settings: unemployed young people who were stuck inside using PlayStation until *Pokémon Go* reintroduced them to the outside world.

Each of the *Pokémon* players has their own story to tell about how the game has helped them. One retired grandmother in the group first took up the game to do something with her grandkids that they might enjoy, and it grew from there. She compared it to being in a choir, as many of her contemporaries are. 'You go to a raid and you do your part. The advantage is that you can either talk to others or just stand there quietly.'

Another of Lovisa's fellow players has a partner with cancer. The game is a chance for him to get out and think about something else for a short while. Several have suffered from long-term depression and take joy in helping newcomers get started in the game. Lovisa's new best friend from *Pokémon*, Cecilia, has Asperger syndrome and ADHD, a symptom of which is an urge to want to hoard things, such as receipts and magazines. 'Now I can collect and organize without becoming a hoarder. And get exercise at the same time!' she told Lovisa. Cecilia has a blunt honesty and humour that helps Lovisa process her own feelings.

Pokémon Go brings stability to Lovisa's life, and to the lives of many others. The rewards arrive in a steady flow, albeit at unpredictable times, but they keep coming. 'It isn't a cure. It is about managing the symptoms,' Lovisa has told me, 'it's a survival mechanism.'

Lovisa and her friends are just one group of many *Pokémon* friends spread across the planet whose lives have been improved by walking around and collecting rewards. Yennie Solheim Fuller, who is senior manager responsible for civic and social impact at Niantic, told me about one player she'd met who was suffering from post-traumatic stress disorder after coming back from an overseas tour of duty: 'Making progress in the game forced him to get out of the house and allowed him to focus on something else than the PTSD.' 'Another big group is the autistic community,' Yennie said. 'After our *Pokémon Go* lunches, we meet so many parents whose kids have had such incredible sensitivity to noise and chaos around them and haven't been able to go outside. Now they are standing in front of the art school doing raids and talking to different people.'

Yennie has received messages from cancer patients thanking them for the game that has got them through tough times. She read out one letter from the son of a man who had had diabetes for fifteen years and had started playing *Pokémon Go*. 'He has now reached level forty. The highest level,' his son wrote, 'and has become one of the most empathic of senior citizen players. The diabetes no longer threatens his health so he doesn't have to do injections.'

This was just one of many stories that made Yennie and her colleagues cry, and, as she read them out to me, I started to cry too. Lovisa reached level forty in summer 2018. And while, to the external world, this might not appear to be one of her most impressive achievements, for me it is a demonstration of how she has used rewards to help her handle her pain.

*

Herbert Robbins' and Sutton Monro's result was a starting point for the branch of mathematics used for signal detection that took off during the 1950s and 1960s. They had shown that a tracking variable, Q_t, could be used to assess changes in our environment. Rewards, good and bad, could be monitored. In 1960, Rudolf E. Kálmán published a seminal paper showing how noise in rewards could be reliably filtered out to reveal the true signal.[7] His technique was used to estimate the speed and position of objects and resistance in rotors,[8] an essential step towards the development of automatic sensors.

The theory of signal detection was then combined with the emerging field of mathematical control theory. Irmgard Flügge-Lotz had already developed the theory of bang-bang automatic control, which provided an automated way to give on–off responses to changes in temperature or air turbulence.[9] Her work, along with that of other control theorists, allowed engineers to design automated systems that monitored and responded to changes in the environment. The first applications were in thermostats that regulate the temperature in our fridges and in our homes. The same equations became the basis of cruise control in aeroplanes. They were also used to align the mirrors within powerful telescopes looking deep into our universe. It was this type of maths that controlled the thruster that performed the initial stages of braking as the *Apollo 11* lunar module approached the moon. Today, it is used inside robots working on Tesla's and BMW's production lines.

Control theory created a world of stable solutions. Engineers wrote down equations and demanded that the world follow their rules. For many applications this worked fine. But the world isn't stable: there are fluctuations and random, arbitrary events.

As the 1960s ended with a new counterculture that challenged the established order, TEN also underwent a revolution. The focus moved from stable, linear engineering to the unstable, chaotic and non-linear. It was this mathematics that influenced me as a young PhD student in the late 1990s and I set about learning it all: mathematical theories with exotic names like the butterfly of chaos, sandpile avalanche models, critical forest fires, saddle-node bifurcations, self-organization, power laws, tipping points . . . Each new model helped to explain the complexity we saw all around us.

A key insight was that stability was not always desirable. The new mathematical models expressed how ecological and societal systems changed – not always returning back to the same stable state again, but sometimes tipping between states. They described how ants formed trails to food, how neurons fired in synchrony, how fish swam in schools and how ecological species interacted. They described how humans make decisions, both in terms of the internal processes in our brains and how we negotiate decisions in groups. As a result of these insights, TEN's members were able to take up positions in biology, chemistry and physiology departments.

This was the mathematics I applied to the data collected by the biologists I worked with.

*

I have a whole range of apps – not just Twitter – that I open and update on my phone. Similarly, ants and bees don't have just one source of food but many alternatives they could choose to visit. There are lots of levers on the slot machine and we don't have time to pull them all. The dilemma is which lever to pull. We know that if we just pull one lever we can get a pretty good idea of the available rewards for that slot machine. But if we spend all our time pulling that lever we'll never find out what the other machines have to offer. This is known as the exploitation v. exploration dilemma. How much time should we spend exploiting what we know versus spending time exploring less familiar alternatives?

Ants use the chemical pheromone to solve this problem. Remember that the amount of pheromone reflects the quality estimate Q_t that ants have in a food source. Now imagine that the ants have two food sources with different pheromone trails to each of them. In order to choose which trail to follow, each ant compares the amount of pheromone on the two trails. The more pheromone there is on the trail, the greater the probability that the ant will follow that trail.

The choices of each successive ant result in a process of reinforcement: the more ants take a particular trail and get a reward, the more likely their nest-mates are to follow them. Trails with lots of previous traffic are reinforced; others are forgotten. This observation can be formulated in terms of Equation 8, with an additional factor that includes the choices of the ants.[10] One example is:

$$Q_{t+1} = (1 - \alpha)Q_t + \alpha \left(\frac{(Q_t + \beta)^2}{(Q_t + \beta)^2 + (Q'_t + \beta)^2} \right) R_t$$

The new term indicates how the ant chooses between two alternative trails. Q_t can be thought of as the pheromone on the trail leading to one potential food source, while Q'_t is the pheromone leading to an alternative food source. We now have two tracking variables (Q_t and Q'_t), one for each food source, or, if we are modelling social media usage, one for each app on my phone.[11]

When faced with a new and complicated equation with lots of parameters, the trick is always to first consider a simpler version. Let's look at the new term without the squares, i.e.

$$\frac{Q_t + \beta}{Q_t + \beta + Q'_t + \beta}$$

As $\beta = 0$, then this is simply the ratio of the two tracking variables. The probability that an ant exploits a particular reward is proportional to the tracking variable for that reward. Now consider what happens if $\beta = 100$. Since Q_t is always between 0 and 1, it is very small in comparison to 100, so the ratio above becomes approximately equal to $100/(100 + 100) = 1/2$. The probability that the ants exploit a particular reward is random, fifty–fifty.

The problem of balancing exploration and exploitation becomes a problem of finding the optimal level of trail reinforcement. This is the same as the problem of determining the correct value of β. Reinforcement that is too strong (β very small) – always following the strongest trail – means the ants always follow the trail with the most pheromone. Very soon none of the ants are visiting the other food source and, even if it improves, none of them will know about it. As a result, the ants become locked into the food source that appeared to be best initially, even if the quality changes later on. Reinforcement that is too weak (β very large) leads to the opposite problem. The ants wander randomly down trails and don't benefit from their knowledge about which is best.

The answer to the explore/exploit problem involves an unexpected twist. It turns out that solving the optimal reinforcement dilemma is related to another concept that usually arises in an entirely different context: that of tipping points.

Let me explain. Tipping points occur when there is a critical mass, and the system tips from one state to another, for example, a fashion suddenly takes off after influencers push a brand or a riot starts as a small group of agitators rouse the protestors.[12] In each of these examples, and many more, reinforcement between people's beliefs leads to sudden changes of state. A similar reinforcement can be seen in ants. We can think of a formation of a pheromone trail as occurring when a tipping point is reached: a trail starts as a small group of ants decide to take the same route to food.

And here is the surprising conclusion: the best way to balance exploration and exploitation is for the ants to remain as close as possible to their tipping point. If the ants go too far past their tipping point, too many ants will commit to one food source and they will become 'locked' into that food and unable to switch if something better comes along. On the other hand, if not enough ants commit to the food source, and they don't reach the tipping point, then they never focus on the best food. The ants have to find that sweet spot, the tipping point, between exploration and exploitation.

Ants have evolved to make sure they remain at the tipping point. One of my favourite examples of ants achieving this balance was discovered by biologist Audrey Dussutour in a species known as big-headed ants (on account of their unusually large heads). These ants have a lot to be big-headed about: they have colonized much of the tropical and subtropical world, out-competing other native species. Audrey found that they deposit two pheromones: one that evaporates slowly and produces weak reinforcement and another fast-evaporating pheromone that produces a very strong reinforcement.[13]

Fellow mathematician Stam Nicolis and I developed a model with two reward equations: one for the weak but long-lasting pheromone and one for the strong but short-lived pheromone. We showed that the combination of the two pheromones allowed the ants to remain near to the tipping point. In our model, the ants were able to keep track of two different food sources, switching between them whenever the quality of the food changed. Audrey confirmed our predictions in experiments: whenever she changed the quality of the food sources, the big-headed ants were able to tip their foraging towards the best quality food source.

It isn't just ants that live life at the tipping point. For many animals, life is an endless casino of slot-machine handle-pulls. Is there a predator in that bush? Is there food where I found it yesterday? Where can we find shelter for the night? To survive in these environments, evolution has brought them to tipping points. This was the phenomenon I discovered time and time again during the fifteen years I studied animal behaviour: marching locusts form at densities that allow them to quickly switch direction; fish schools suddenly expand as a shark attacks; starling flocks turn in unison away from a hawk. By moving together, the prey en masse confuses the predator.

Animals have evolved to be close to a tipping point. They are in a state of constant collective awareness – switching from one solution to the next; highly responsive to changes. For them it is a matter of survival.

But for humans? Are we stuck at a tipping point? And, if so, should we be there?

*

In 2016, Tristan Harris struck out at social media. For the previous three years he had worked as a design ethicist at Google, but now, it seemed, he had had enough. He left Google and went to the online publishing platform Medium, where he posted his manifesto. The title claimed that 'technology is hijacking your mind' and the twelve-minute text explained how they were doing it.[14]

The analogy Harris chose for social media should be familiar by now: the slot machine. He described the technology giants as having put slot machines in several billion people's pockets. Notifications, tweets, emails, Instagram feeds and Tinder swipes were all asking us to 'pull a lever' and find out if we had won. They interrupted our day with continual reminders, then drew us in through a fear of missing out if we didn't pull the lever. They enticed us with the social approval of our friends and encouraged us to reciprocate by liking and sharing as we pulled the levers together. All of this was done according to the tech companies' own agendas – to get you to watch ads or follow sponsored links. Google, Apple and Facebook had created a gigantic online casino and were raking in the profits. The reason that our pocket slot machines are so addictive is that they place us in a continual dilemma between exploring and exploiting. And social media isn't any ordinary slot machine: it has thousands of handles that all need to be pulled in order to know what is going on.

Scientists have long known about the difficulties multiple handles pose for the brains of animals. In 1978, John Krebs and Alex Kacelnik performed an experiment on great tits in Oxfordshire.[15] They provided the birds with two different perches. The perches were constructed so that when the tits hopped on either perch, food would sometimes drop down for the bird to eat. The probability of food dropping down differed between the two perches so that one was more likely to drop down food than the other. What Krebs and Kacelnik found was that, when one perch was a lot more profitable than the other, the birds quickly focused on exploiting that perch. But when the perches were quite similar, they found the task difficult. They would move backwards and forwards between the two perches, testing out which was best. In my terminology, the birds came close to the tipping point.

Mathematician Peter Taylor has shown that the reward equation is entirely consistent with this result. The more difficult it is to decide between rewards, the more exploration is required. We do the same as the great tits, but with even more choices. We open app after app. But it isn't the availability of all of these rewards that is the problem, it is our brain's desire to explore and exploit. We want to make sure we know where each of the potential rewards can be found. We are pushed towards the 'edge of tipping'.

There is a big difference between exploiting only one source of reward and many. When you read a book, play *Mario Kart* or *Pokémon Go*, binge-watch *Game of Thrones*, have a game of tennis with a friend or go to the gym, you are focusing on just one source of reward. You enjoy the repetitive item-collection dings and lap-completed jingles.

Your reward equation converges to a stable state. This is the reward equation of the 1950s, for which Robbins and Monro proved the steady convergence. You learn what to expect from the activity and slowly but surely your confidence matches the rewards it provides. It is this familiar stability that brings you pleasure.

When you are using social media, on the other hand, you are exploring and exploiting lots of different sources of reward. In fact, you aren't really collecting rewards at all: you are monitoring an uncertain environment. Remember, dopamine isn't a reward and so you aren't getting pleasure. You are in survival mode, gathering as much information as possible. It isn't necessarily the unlimited availability of rewards that is the problem; it is your need to monitor all the different potential sources of reward that makes your life difficult. You are putting your brain at a tipping point – at the edge of chaos, at a phase transition. No wonder you are stressed.

And it isn't just your brain that is at the tipping point: our entire society is there. We are like ants running around manically, trying to keep track of all the information sources. And these information sources are constantly moving around, changing quality and sometimes disappearing completely. So, what can you do about this problem?

The organization Tristan Harris co-founded, the Center for Humane Technology, has some advice about how to take control and move your own mind away from the tipping point. You should turn off all the notifications on your phone so that you aren't disturbed by constant interruptions. You should change your screen settings to make your phone icons less colourful so they don't catch your eye so easily.

For the most part I'd agree with Harris's advice. It is common sense. But there are also some less obvious, and possibly even more useful, insights to be gained from the way that ants use the reward equation.

First, you should recognize the incredible power of having your mind and society as a whole at a tipping point. It is not a coincidence that the most ecologically successful ant species are the ones which use pheromone most effectively. The same applies to humans' transition to the tipping point. While it might stress you as an individual, a society on the verge of the tipping point is able to produce and spread new ideas more quickly. Think about the wealth of ideas that has come out of the #MeToo movement or #BlackLivesMatter. These campaigns really have made people aware of issues and can produce change. Or if you are of a different political persuasion, look at Trump's election or #MakeAmericaGreatAgain. Think about how these ideas took off and the response to them, from both sides.

We are more engaged nowadays in political debate. When it comes to political causes, young people are more active than ever before, both online and in real life.[16] We are like a flock of birds whirling in a murmuration in the evening sky. We are a school of fish rotating as a predator approaches. We are a bee swarm flying to a new home. We are a colony of ants raiding the forest floor for food. We are a human crowd surfing on the news.

Take yourself to that tipping point and enjoy the freedom of being there. Click through news article after news article. Get information, take in new ideas and follow your interests. During the writing of this book, I have 'wasted' uncountable hours on Google Scholar, browsing scientific articles, looking to see who has cited whom and deciding which are the important scientific questions. Chat with people online. Argue if necessary. Send an email to some daft old professor who is writing for *Quillette*. Engage, be part of the information flow. Then, once you have spent an hour or so at the tipping point, I can tell you about the second insight I gleaned from looking at the ants.

I realize that I might have given the impression of ants as hyperactive slot-machine junkies. This is true when they are at work and some ants do work very hard. But many of them can be very lazy indeed. At any given time, the majority of ants are doing absolutely nothing.[17] While a minority are running around like crazy, evaluating and collecting food, most of their nest-mates are simply resting. Some of this inactivity is associated with working in shifts; not all ants are active at the same time. But colonies also contain a lot of ants who do very little, hardly ever go outside and don't tidy up either. Why ants have evolved to allow such lethargy no one knows for sure, but if we are going to admire the minority for their high levels of activity, then we should also give the majority credit for their laid-back attitude to life.

So, when you have been at the tipping point for a while, live life like a lazy ant. Switch off. Get *Game of Thrones* going on autoplay. Watch *Friends* again from the first episode to the last. Spend a whole week or month collecting all the Pokémons. Of course, I should add all those morally superior activities like going for a walk, sitting on the porch and fishing. But the main thing is you should relax – without your phone. Pay no attention to the news and ignore the endless round of cc'd emails. Don't worry, someone else will take care of it. It doesn't have to be about you all the time. We have got this.

The reward equation tells you to concentrate on the present and not to dwell on the past. Keep track of where you are using a single number in your head. When things work out, update the number; when they don't, let your estimate evaporate a tiny amount. Make sure you spot the difference between stable rewards that keep on giving (albeit sporadically) no matter what you do and unstable rewards that change their nature over time. Stable rewards can be found in friendships and relationships, books, films and television, long walks and fishing, 2048 and *Pokémon Go*. Unstable rewards are found on social media, looking for a partner on Tinder, in most jobs and often (whether we like to admit it or not) within our family life. Don't be scared to explore and exploit in these situations, but remember that you are getting the most from these rewards when you are at the tipping point. So, before the unstable rewards tip you somewhere you don't want to be, find your way back to stability again.

9

The Learning Equation

$$-\frac{d(y - y_\theta)^2}{d\theta}$$

You've probably heard that technology in the future will be dominated by artificial intelligence (AI). Already, researchers have trained computers to win at Go and self-driving cars are being tested out. It is one thing for me to explain a small number of equations in this book, but surely I have forgotten something, haven't I? Don't I also need to tell you the secrets behind the AI used by Google and Facebook? Shouldn't I tell you how we can get computers to learn to think like us?

I'm going to let you into a secret, one that doesn't quite fit with the narrative of films such as *Her* or *Ex Machina*. Neither does it tally with the concerns expressed by Stephen Hawking or the hype of Elon Musk. Tony Stark, fictional superhero Iron Man of the Marvel Comics, wouldn't be happy with what I am about to say: artificial intelligence in its current form is no more (or less) than the Ten Equations put together by engineers in an imaginative way. But before I explain how AI works, it is time to pause for a commercial break.

*

Around about the time of 'Gangnam Style', YouTube had a problem. It was 2012 and although hundreds of millions of us clicked on videos and visited the site, we weren't staying there. Novelty videos like 'Charlie Bit My Finger', 'Double Rainbow', 'What Does the Fox Say?' and 'Ice Bucket Challenge' only held our attention for thirty seconds before we went back to watching TV or doing something else. In order to attract advertising revenue, YouTube needed to become a place where users would stick around.

YouTube's algorithm was a big part of the problem. It was using a system to recommend videos based around the advertising equation we saw in Chapter 7. A correlation matrix was built for videos that users watched and liked. But this method didn't account for the fact that young people wanted to see the most recent videos. Nor did it take on board how engaged users were with a video. It simply showed the videos that other people had watched. As a result, the Norwegian army doing the 'Harlem Shake' just kept popping up again and again in recommended videos lists and users kept disappearing from the site.

YouTube called the Google engineers. 'Hey Google, how can we help kids find the videos they like?' they asked (probably). The three engineers – Paul Covington, Jay Adams and Emre Sargin – who were assigned to the task soon realized that the most important criterion for YouTube to optimize was watch-time. If YouTube could get its users to watch as many videos as possible for as long as possible then it could insert adverts at regular intervals and make more money. Shorter, novelty videos were thus less important than YouTubers who set up channels, providing a continuous supply of fresh, longer-form content. The challenge was to find a way of identifying this content on a platform where hours and hours of video are uploaded every second.[1]

The engineers' answer came in the form of a 'funnel'. This was a device that took hundreds of millions of film clips and reduced them down to a dozen or so recommendations presented in a side bar on YouTube. Every user was given their own personalized funnel which found the videos they would most likely want to watch.

The Funnel is a neural network, a series of interconnected neurons which learns about our watching preferences. Neural networks can best be visualized as a column of input neurons on the left and a column of output neurons on the right. In between there are layers of connecting neurons, known as hidden neurons (see Figure 9). There may be tens or even hundreds of thousands of neurons in a neural network. These networks aren't real in a physical sense: they are computer codes which simulate the interactions of the neurons. But the analogy to the brain is useful, because it is the strength of the connections between the neurons which allows neural networks to learn about our preferences.

Each neuron encodes aspects of how the network responds when presented with input data. In the Funnel the neurons capture relationships between different items of YouTube content and channels. For example, people who watch right-wing commentator Ben Shapiro also tend to watch Jordan Peterson videos. I know this because, after completing my research for Chapter 3 on the confidence equation, YouTube became obsessed with offering me Shapiro videos. Somewhere inside the Funnel there is a neuron that represents a connection between these two 'Intellectual Dark Web' icons. When the neuron receives an input telling it that I am interested in Peterson videos, it outputs that I might also be interested in Shapiro videos.

We can understand how artificial neurons 'learn' by studying the way connections are formed inside the network. Neurons encode relationships in terms of parameters – adjustable values that measure the strength of relationships. Let's consider the neuron responsible for working out how long users will spend watching Ben Shapiro videos. Inside this neuron there is a parameter called θ which relates time spent watching Shapiro to the number of Jordan Peterson videos a person has watched. For example, we might predict that the number of minutes a user spends watching a Shapiro video, denoted y_θ, is equal to θ multiplied by the number of Peterson videos a person has watched. So, if $\theta = 0.2$, then a person who has watched ten Peterson videos is predicted to watch a Shapiro video for $y_\theta = 0.2 \cdot 10 = 2$ minutes. If $\theta = 2$, then the same person is predicted to watch for $y_\theta = 2 \cdot 10 = 20$ minutes of Shapiro, and so on. The learning process involves tuning the parameter θ to improve predictions of viewing time.

Let's assume that the neuron's initial setting is $\theta = 0.2$. I come along, having watched ten Peterson videos, and end up watching Shapiro for $y = 5$ minutes. The squared difference between prediction (y_θ) and reality (y) is thus

$$\left(y - y_\theta\right)^2 = \left(5 - 2\right)^2 = 3^2 = 9$$

We have seen this idea of squaring differences before, in Chapter 3, when measuring standard deviation. By calculating $(y - y_\theta)^2$ we get a measure of how good (or bad) the neural network's predictions are. The discrepancy between prediction and reality of 9 is big, so the prediction wasn't that good.

In order to learn, the artificial neuron needs to know what it did wrong when it predicted I would only watch for two minutes. Since the parameter θ controls the strength of the relationship between the number of Peterson videos watched and the typical time a user might spend watching Shapiro, increasing θ also increases the predicted watch-time y_θ. So, for example, if we make a small increase of $d\theta = 0.1$ to θ, we get $y_{\theta+d\theta} = (\theta + d\theta) \cdot 10 = (0.2 + 0.1) \cdot 10 = 3$ minutes. This prediction is closer to reality:

$$(y - y_{\theta+d\theta})^2 = (5 - 3)^2 = 2^2 = 4$$

It is this improvement which Equation 9, the learning equation, exploits.

$$-\frac{d(y - y_\theta)^2}{d\theta}$$

(Equation 9)

This expression says we should look at the way a small change $d\theta$ to θ increases or decreases the squared distance $(y - y_\theta)^2$. Specifically, in our example:

$$-\frac{d(y - y_\theta)^2}{d\theta} = -\frac{(y - y_{\theta+d\theta})^2 - (y - y_\theta)^2}{d\theta} = -\frac{4 - 9}{0.1} = 50$$

The positive value 50 means that increasing θ improves the quality of the prediction – and the distance between prediction and reality decreases.

The mathematical quantity calculated by Equation 9 is known as the derivative or the gradient with respect to θ. It measures whether a change θ brings us nearer or further away from good predictions. The process of slowly updating θ, based on a gradient, is often referred to as gradient ascent, invoking an image of us climbing up a hill by following the slope upwards. By following the gradient, we can slowly improve the accuracy of the artificial neuron (Figure 9).

The Funnel doesn't just work on one neuron at a time – it works on all of them. Initially, all the parameters are set to random values and

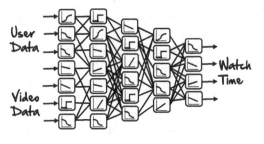

A section of the Funnel neural network. Each neuron is a function that takes in inputs and outputs a prediction.

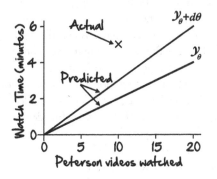

Inside each neuron the function is tuned to make better predictions. Here, an increase $d\theta$ in θ moves the predicted watch time closer to the actual watch time.

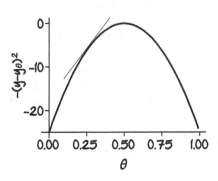

By climbing up the gradient, towards the point at which the distance between prediction and observation can no longer be improved, the neurons 'learn' relationships in the data.

Figure 9: How a neural network learns

the neural network makes very poor predictions about the length of time people will watch videos. Then the engineers start feeding the watching patterns of YouTube users into the input neurons at the wide end of the Funnel. The small number of output neurons (those at the narrow tip on the right-hand side of the Funnel) measure how well the network predicts the length of time people watch videos for. At first, the errors in predictions are very large. Through a process called backpropagation, errors in prediction measured at the tip of the network feed backwards through the Funnel's layers. Each neuron measures the gradient and improves its parameters. Slowly but surely the neurons ascend the gradient and the predictions gradually improve. The more YouTube user data is fed into the network, the better the predictions become.

The Shapiro/Peterson neuron, from my example above, is not encoded inside the network from the start. Indeed, the power of neural networks is that we don't need to tell them which relationships to look for in the data, the network finds these relationships through the process of gradient ascent. Because the relationship between Shapiro/Peterson predicts viewing times, eventually one neuron, or a small number of them, will start to exploit that relationship. These neurons will interact closely with other neurons related to other 'Intellectual Dark Web' celebrities and even to more extreme right-wing ideologies. It creates a statistically correct representation of the type of person who is likely to watch Jordan Peterson videos.

Equation 9 is the basis of a set of techniques known as machine learning. The gradual improvement of parameters using gradient ascent can be viewed as a process of 'learning': the neural network (the 'machine') gradually 'learns' to make better and better predictions. Provided with enough data – and YouTube has lots of it – the neural network learns the relationships within that data. Once it has 'learnt', the Funnel can predict how long a YouTube user will spend watching videos. YouTube puts this technique into action. It took the videos with the longest predicted viewing times and put them into users' recommended lists. If the user didn't choose a new video, YouTube would simply autoplay the video it thought he or she would like the best.

The success of the Funnel was astounding. In 2015, the time eighteen-to forty-nine-year-olds spent watching YouTube increased by 74%.[2] By 2019, it had twenty times as many views as it did before the Google researchers started their project, with 70% of those views coming from recommended videos.[3] Doug Cohen, Snapchat data scientist, was full of admiration for the solution. 'Google solved the explore and exploit problem for us,' he told me. Instead of having to click around different sites and try to find the best videos, or wait until someone sends you an interesting link, now you could sit for hours in front of YouTube either watching the 'Up Next' videos or clicking on one of the ten 'alternatives' you were being offered.

If you believe that you are exploring your own interests on You-Tube, but find yourself clicking on the suggested videos, then you are sadly misguided. The Funnel has effectively turned YouTube back into a traditional form of TV, with scheduling decided by AI. And many of us have become glued to its screen.

*

Noah would like to be more popular on Instagram. Lots of his friends have more followers than he does, and he watches with envy as they accumulate 'likes' and 'comments' from others. He looks at his friend Logan's account: he has around 1,000 followers and every post he makes gets hundreds of likes. Noah wants to be like Logan and sets his own goal to have $y = 1,000$ followers. With his current social media strategy he has only $y_\theta = 137$. There's a long way to go.

Over the course of the following week Noah gradually starts to put up more posts. He reckons that the more he posts, the more people will follow him. He takes pictures of his dinner, of his new shoes and his walk to school, but he doesn't make any effort to improve the quality of his pictures. He is just photographing everything he sees and putting it out there. In terms of Equation 9, the parameter θ which Noah is adjusting is the ratio of quantity to the quality of his posts. He is increasing the quantity of posts and $d\theta > 0$.

The online response isn't good. 'Why u spam us,' writes his friend Emma under one of his pictures, adding a puzzled looking emoji to her text. Some of Noah's acquaintances stop following him. His popularity has gone down: $y_{\theta+d\theta} = 123$, a decrease of 14. The distance between this number and his target has increased. He is descending the gradient instead of climbing it. Over the coming months Noah tunes the quantity down and focuses on quality instead. A couple of times a week he takes a picture of a friend enjoying an ice cream or a funny picture of his dog. He makes sure he carefully edits the picture and uses a filter that makes his friend look good. As he makes the switch from quantity to quality, he now measures y_θ. The number of his followers has increased slowly but surely. After six months the number of followers has climbed to 371, but then it stabilizes and he doesn't gain any more followers during the seventh month.

We now come to the important lesson from Equation 9: Noah should relax and stop trying to reach his goal of 1,000 followers. Despite the fact that $(y - y_\theta)^2 = (1,000 - 371)^2 = 395,641$ is still very big, Equation 9 is no longer changing:

$$-\frac{d\left(y - y_\theta\right)^2}{d\theta} = 0$$

The equation tells Noah to stop playing about with his social media strategy and be satisfied with what he has got. There is no longer any need to compare himself with Logan: Noah has reached his own popularity peak.

When applying Equation 9 we should keep our overall target in mind, but be guided primarily by whether or not we are moving upwards. As every wise old woman knows, when you are at the top of the mountain, you should enjoy the view. The maths supports this traditional wisdom.

The difference between the type of optimization that machine-learning algorithms like the Funnel create and the optimization done by Noah is that, while he was trying to increase his number of followers, machine learning is attempting to optimize the accuracy of its predictions. In the case of the Funnel, y_θ are predictions of how long users will watch videos for and y is how long they actually spend watching them. YouTube wants to be able to predict its users' preferences as accurately as it possibly can, but it realizes that its predictions will never be perfect. The Funnel is satisfied when it realizes that it can't get better.

The trick to using the learning equation is to be honest about how your actions increase or decrease the difference between your goals and reality. Some people might accuse Noah of being 'fake' or 'superficial' when he attempts to actively optimize his social media impact. I don't agree. Working with behind-the-scenes influencers like Kristian Icho, a colleague of mine who runs a social media site about street fashion, has taught me otherwise. Kristian uses Google's data analytics tool to study how the ratio of quality to quantity in posts produces customer flow, but he still understands that his data is about people. When a seventeen-year-old kid posted a selfie in a designer T-shirt, Kristian's face would light up. He'd give the post a 'like' and comment, 'Looking good!' He meant it. There is no contradiction between learning from data and being 100% genuine about who you are and what you do.

Used carefully, the learning equation does help you optimize your own life. Whether you are trying to make it on social media or study for your exams, always aim to climb the gradient slowly. Set yourself a target, but don't focus on the distance between yourself and your goal. Don't worry about the people who are more popular than you are or your peers who get better grades. Focus instead on the steps you are taking each day. Focus on the gradient: the friendships you gain or the new understanding you reach in your studies. If you find you aren't making progress, then admit this to yourself. You have reached the hilltop and it is time to enjoy the view for a while. Be aware, though, that following the gradient is by no means a perfect technique; sometimes you get trapped in a suboptimal solution. Then it is time to reset and start again. Find a new mountain to climb or a new parameter to adjust.

*

In 2019, Jarvis Johnson quit his full-time job as a software engineer. His YouTube channel, where he made videos about his life as a programmer, was attracting more and more subscribers. He decided to see if he could make it as a full-time 'Internet person', as he now calls himself.

Making it as a YouTuber requires two things: posts with interesting content and a deep understanding of the Funnel process. Jarvis has both and his videos combine these elements with a self-referential humour. He investigates how some YouTube channels are exploiting the algorithm for their own benefit by manipulating the Funnel so that it points all its recommendations towards them. Then he turns his findings into entertaining and engaging videos on the platform itself.

Jarvis's investigations have focused on a publishing group called TheSoul Publishing. They describe themselves as 'one of the largest media publishers in the world' and claim that they are on a mission to 'engage, inspire, entertain and enlighten'. Jarvis started by looking at one of TheSoul's most successful channels: 5-Minute Crafts. It purports to offer 'Life Hacks', tips to make your everyday chores easier. In one video, with 179 million views, the site claimed that permanent marker could be cleaned from a T-shirt using a combination of hand sanitizer, baking powder, lemon juice and a toothbrush. Jarvis decided to test this out himself, writing the word NERD across the front of his white T-shirt and then following the cleaning instructions. The result? Instructions followed and even after machine-washing, NERD remained there in bold permanent letters. Hack after hack, Jarvis tried them out and showed that they were either trivial or didn't work. The tips offered on 5-Minute Crafts were useless.

One of TheSoul's other channels, Actually Happened, claims to animate real-life stories from its followers. Jarvis found that its content was created by 'scriptwriters' who made up believable stories that would appeal to a US teen audience, using Reddit and other social media sites as a source. Jarvis explained to me that Actually Happened was initially copying another channel, Storybooth, which does animate genuine personal stories from children and teenagers. The latter channel often has the kids themselves narrate their story, which helps to present it in a way that is honest and genuine.

'The YouTube algorithm can't tell the difference between Storybooth and Actually Happened,' Jarvis told me in an interview in May 2019. The Actually Happened channel uses the same titles, descriptions and tags as Storybooth and the YouTube Funnel views them as more or less equivalent and starts making links between the two sites. 'Actually Happened flooded the market with stories. They hired in contractors below the market rate and they pumped out a video a day,' Jarvis went on, 'then they reached an escape velocity where they don't have to copy Storybooth anymore.' Once they had over 1 million subscribers, the Funnel decided that Actually Happened was a channel that kids wanted to watch. The algorithm saw it as a hilltop for its users.

Jarvis thought that the morality issues around TheSoul's channels were complicated. 'I've certainly made content similar to others with the hope of sharing an audience, but doing it en masse and so overtly is something that I couldn't do. What's to stop this company from doing this with every genre on the platform?'

The limitation of the YouTube algorithm is that it doesn't care about the content it promotes or the work that has gone into making it. I experienced this myself when YouTube decided I was interested in Ben Shapiro. Anyone who has put their young children in front of YouTube for an hour or so will very likely have experienced how they are sucked into a strange world of toy-unwrapping videos, colourful Play-Doh ice-cream cups and Disney wrong-head puzzles. Check out 'PJ Masks Wrong Heads for Learning Colors': it looks as if it took about half an hour to make and now has 200 million views. The videos recommended by the Funnel are not just poor quality, they can also be highly inappropriate. In 2018, *Wired* magazine documented videos in which the PAW Patrol dogs attempted suicide and Peppa Pig was tricked into eating bacon.[4] One *New York Times* investigation found that YouTube was recommending family videos of naked children playing in paddling pools to its users who took an interest in paedophilia.[5]

YouTube can give us 'funnel vision'. Its target might be to reach the absolute best recommendations for you, but Equation 9 is satisfied when it finds the best solution given the data it has available. It climbs the learning gradient until it reaches a peak and then it stops to let you enjoy the view, whatever that might happen to be. The Funnel makes mistakes and it is our responsibility to keep setting it straight. YouTube hasn't always succeeded in rising to this challenge.

*

From the outside, some may envisage TEN's members as being like Tony Stark's Iron Man: industrialists and talented engineers who are using technology to transform the world. But if each of the members of TEN were to choose a Marvel superhero to describe themselves it would probably be Spider-Man/Peter Parker. They have no plan, no moral agenda – they are just like teenagers trying to maintain control over their bodies as they develop in unexpected ways.

The tension within TEN's members can be viewed in many ways. Are its members like the naive Mark Zuckerberg from the film *The Social Network*, or the robot-like Mark Zuckerberg giving his testimony to the US Senate Judiciary and Commerce committees? Are they like the Elon Musk who smokes dope on TV or the one who thinks our future lies with a move to Mars?

On the one hand, the equations give TEN impeccable judgement so that they are trusted with planning global changes to our society. They have created a scientific approach that builds confidence in models through data. They have connected us all in ways we never expected. They optimize and improve performance. They bring efficiency and stability. On the other hand, members will adhere to a reward equation that tells them to take what they can now and forget about the past. They create edges over those who can't afford to pay.

This is exactly what A. J. Ayer told us in 1936: there is no morality in maths, or, if there was once, it has now been lost. TEN's invisibility has meant that we can't even find the proper superhero analogy. Are TEN's members just naive teenagers, realizing like Peter Parker that with great power comes great responsibility, or are they power-crazy maniacs who want to control the world 'for its own good'? Might they even be like Marvel's super-villain Thanos, ready to kill half the population because they think it is the optimal thing to do?

Whatever they think they are, we need to be aware of what they are up to, because, wherever they go, they change everything.

*

When we look at an example of modern artificial intelligence (AI), such as Google's DeepMind neural network which learnt to be the world's best Go player or the AI that learnt to play *Space Invaders* and other Atari games, we should see them as feats of engineering. A team of mathematicians and computer scientists have put all the pieces together. There isn't one single equation which is behind this AI.

But – and this is important for my whole project of revealing the Ten Equations – the components of artificial intelligence involve *nine* of the Ten Equations. So now for my final trick I am going to explain how DeepMind became a master games player using the maths we have learnt so far in this book.

Imagine a scene where a grandmaster of chess is standing in the middle of a ring of tables. He goes up to one table, studies the board and plays a move. He then moves on to the next table and plays that one. At the end of the challenge he has won all of his matches. Initially it might seem incredible that the grandmaster can keep track of so many games of chess at the same time. How can he remember the way in which each game has progressed up to this point and decide what to do next? But then you recall the skill equation.

The state of a game of chess can be seen immediately from the board: the defensive arrangement of the pawns, how well the king is hidden, how free the queen is to attack and so on. The grandmaster doesn't need to know how the game has gone up to that point, he just needs to look at the state of the board and decide his next move. The skill of the grandmaster can be measured by the way he takes the current state of the board and converts it into a new state, via a valid move. Does that new state increase or decrease the chance of his winning the game? When assessing grandmasters, Equation 4 (the Markov assumption) applies.

'Many games of perfect information, such as chess, checkers, othello, backgammon and Go, may be defined as alternating Markov games.' This was the first sentence in the Methods section of a paper by David Silver and other Google DeepMind researchers about their world-champion-beating Go neural network.[6] This observation immediately made the problem of solving these games simpler, by allowing them to focus on finding the best strategy for the current state of the board without worrying about what happened up to that point.

We have already analysed the mathematics of a single neuron in Chapter 1. Equation 1 took the current odds for a football match and transformed them into a decision about whether or not we should place a bet. This is essentially a simplified model of what a single neuron in your brain does. It takes in external signals – from other neurons and from the outside world – and transforms them into a decision about what it should do. This simplifying assumption was the basis of the first neural network models, and Equation 1 was used to model the responses of neurons. Today it is one of two very similar equations that are used to model neurons inside almost all neural networks.[7]

Next, we turn to a version of the reward equation. In Equation 8, Q_t was our estimate of the quality of a Netflix series or the rewards to be gained by checking a Twitter account. Instead of evaluating just one film or one Twitter account, we now want our neural network to evaluate 1.7×10^{172} different game states in a game of Go or 10^{172} different combinations of YouTube clips and users. We write $Q_t(s_t, a_t)$ to represent the quality of the state of the world s_t given that we intend to carry out a particular action a_t. In Go, the state s_t is represented by a 19 by 19 grid of positions with three states (empty or occupied by white or black). The possible actions a_t are the positions where a stone can be placed. The quality $Q_t(s_t, a_t)$ tells us how good a move we believe action a_t in state s_t to be. For YouTube videos, the state is all the users who are online and the videos available. The action is showing a particular user a particular video and the quality is how long they watch it for.

The reward $R_t(s_t, a_t)$ is the prize we get for carrying out action a_t in state s_t. For Go, the rewards only come at the end of the game. We can say 1 point for a winning move, –1 for a loss and 0 for any other move along the way. Notice that a state can be of high quality but still have zero reward. For example, if it is an arrangement of stones which is close to winning, it has high quality but zero reward.

When DeepMind used the reward equation to play Atari games it added an additional component: the future. When we carry out action a_t (place a stone in Go), we move into a new state s_{t+1} (a board where the site on which the stone was placed is now occupied). The DeepMind reward equation adds a reward of size $Q_t(s_{t+1}, a)$ for the best action in this new state. This provides a way for the AI to plan its future steps through the game.

Equation 8 provides us with a guarantee. It says that if we follow its scheme and update the quality of our play we will gradually learn to play the game. Not only that, by using this equation we will eventually converge on the best overall strategy for any game, from noughts and crosses to chess and Go.

There is a problem, though. The equation doesn't tell us how long we will have to play the game to know the quality of all its different states. The game Go has $3^{19 \times 19}$ states, which is around 1.7×10^{172} different possible board configurations. It takes a long time to play through them all, even with a very fast computer, and in order to allow our quality function to converge we would need to play many times through each state. Finding the best strategy is possible in theory but is impossible in practice.

The key innovation of the Google DeepMind researchers was to realize that the quality $Q_t(s_t, a_t)$ could be represented as a neural network. Instead of trying to learn how an AI player should play in all of the 1.7×10^{172} states of Go, the AI understanding of the game was represented as an input of 19×19 board positions fed into a network, followed by several layers of hidden neurons and output neurons that decided the next move. Once they had formulated the problem as a neural network, the researchers could use gradient descent (Equation 9) to tune it to the answer.

In perhaps the most powerful demonstration of the power of this approach, it took a Google AlphaZero neural network with no experience of chess only four hours to learn how to play to the same level as the best computer chess programs in the world, which were already far ahead of the best human player. From there AlphaZero continued to learn, challenging itself to games in order to find ways of playing that no human, and indeed no other computer, had ever calculated to be possible.

All of the equations we have encountered so far come up in the study of neural networks. We have already used Equations 1, 4, 8 and 9. Equation 5 crops up when we study connections within networks. The way the neurons are connected is the key to determining the type of problem a network can solve. The name 'Funnel' comes from the structure of the neural network that YouTube uses, with a wide entrance of input neurons narrowing down to a small number of output neurons. For other applications, researchers have found that other structures work better. For face-recognition and playing games, a branching structure known as a convolutional neural network works best.[8] For language processing, a network with loops, known as a recursive neural network, is the best option.[9]

Equations 3 and 6 are used when we are looking at the time we need to train a network in order to be confident that it has learnt properly. Equation 7 underlies a method called unsupervised learning which can be used when we have millions of different videos, pictures or texts to analyse and we want to know the most important patterns to pay attention to. Equation 2 forms the basis of the Bayesian neural networks, which are essential in solving games involving uncertainty, such as poker.

So – in just nine equations – we can find the groundwork for modern artificial intelligence. Learn them and you can help create the artificial intelligence of the future.

*

Most of us are unaware that all of the best research in AI is openly available to anyone who would like to learn more and who already understands the nine equations as well as you do now. The articles are published in open access journals and libraries of computer code are available to those who want to start creating their own models. The open secret continues to grow, from de Moivre's doctrine and Gauss's notebooks, through the explosion of science in the last century and now on to the Github archives, where the tech giants upload and share their latest code.

Instead of believing the scare stories or the overblown hype about an artificial intelligence becoming human, the story that people need to be told is about Google. This company, started by two students in California, has driven and financed top-quality research and made almost everything it does available to the public. There are of course dangers with the best minds leaving the universities and going to work for Google, Facebook and the others. But many of us are still here in the ivory towers too, and we are learning (almost) as much from Google today as they have taken from our previous work.

TEN's secrets are not the equations themselves, but in knowing how to apply them and how to combine them. Used without thought, the equations themselves don't solve anything.

The future risk to humanity is not from a hostile AI that takes over our world: a rogue 'Edwin Jarvis' from Marvel's Avengers or the AI 'Samantha' who seduces every man on the planet in the film *Her*. Artificial Intelligence isn't smart enough for that. It gets stuck in its own limited solutions. Rather, the risk lies in the way the gap is widening between those who wield the power over data and those who don't. The small group of humans who know the equations possess an intelligence that has never before been witnessed on our planet.

Mathematically enhanced humans are taking over. Two graduate students created Google Search from the influencer equation (Equation 5). Three Google engineers created a neural network that locked tens of millions of people in to hour after hour of mind-numbing video and advertising viewing. This pattern of having a small number of programmers, financiers or gamblers using maths to dominate everyone else repeats itself over and over again; in other words, a small elite group of mathematicians controlling the lives of those that can't or don't want to learn the code.

Unaccountable for its actions, TEN transforms every aspect of our world. Unconcerned by its limitations, TEN seeks optimal answers to every problem. TEN might well be unaware of itself, but the evidence for its existence is undeniable.

Now we know how nine of the ten equations work, and the strengths and limitations of each of them, we might finally be able to answer the most important question of all. Is this secret mathematical society that runs our world a force for good or evil?

I have personally become richer, smarter and more successful by following TEN, but have I become a better person?

IO

The Universal Equation

If . . . then . . .

I typed the question into my phone: 'Who is better out of Ronaldo and Messi this season?'

I looked up at Ludvig, Olof and Anton, who stood in front of a projection of their laptop screen. Ludvig shifted nervously on his feet. It was his part of the code that would be tested first. Could the Soccer Bot convert my English sentence into a language it could understand?

The text started to roll up the projected screen, the inner workings of the bot brain. My question had become:

{intent: compare; contact: {Ronaldo, Messi}; sentiment: neutral; game span: season}.

The bot had got it! It knew what I meant. Now it was Olof's turn to be nervous. His role had been to model player qualities. I hadn't defined the time period I was interested in, but the bot had used the default of recent matches. Olof's algorithm could classify performances as 'poor', 'average', 'good' and 'excellent'. Now the bot had been asked to assess and compare two excellent players.

The bot decided to go for {weight: shots; tournament: CL} to tell us about the shots made and goals scored in the only tournament they were both playing in, the Champions League. We could see the answer on the projected screens: the bot knew which player it thought was best. What remained now was to send that information back to my phone – not in the form of curly brackets, colons and summarized text, but in a sentence which I could read.

Anton, whose job it had been to construct replies, said, 'There are over 100,000 different things it can say here. Different ways it can put the sentence together and word choices. I wonder which it will choose.'

I looked down at my phone. It took time. We certainly had some work to do with the user interface . . .

Finally it answered, 'Out of these two players, I think Lionel Messi is the best. Lionel Messi has scored six times and has gotten great shooting points, this season.' The bot sent me a link to a shot map with all his shots and goals in the Champions League. It certainly sounded like a bot, with its reference to 'shooting points', but there was also something charming about its phrasing. And, as far as I was concerned, it had got the answer right too.

*

The students' Soccer Bot was built, in part, using the mathematics we have examined in this book. Ludvig used the learning equation to train the bot to understand questions about football. Olof used the skill equation to assess players and the judgement equation to compare them. Then Anton had tied everything together with one final equation: 'If . . . then . . .'.

Before we turn our focus on to this final equation, I want to take a look at where we are in our journey into maths. Let's think a bit about what we have learnt.

Understanding equations can happen on different levels. You can journey into their mathematical depths in order to appreciate exactly how they work and can be used. If your aim is to become a data scientist or statistician working for Snapchat, a basketball franchise or an investment bank, then this journey into technical details is the one you need to make. This book is just a start.

You can also make use of the Ten Equations in a different way – a less technical way, a softer way. You can use them to guide your decision-making and how you think about the world. I believe that you can use the Ten Equations to become a better person.

In Western thinking, the original 'If . . . then . . .' statements were in the Ten Commandments. *If* it is a Sunday, *then* keep it holy. *If* you hear of other gods, *then* none shall come before me. *If* your neighbour's wife is hot, *then* you shall not covet her. And so on. The problem with the Commandments is that they are inflexible, and several millennia later they feel somewhat dated.

The nine equations we have examined so far are different. They don't stipulate rules for what you should and shouldn't do in different situations. Instead, they suggest a way of approaching life. Remember when Amy heard Rachel bitching about her in the bathroom? Or how the friendship paradox helped us see that we were wrongly coveting the social success of others? Or when you stereotyped your friends using the advertising equation? In each of these cases, I didn't tell the actors what they should do based on a predefined moral compass. Instead, I looked at the data, identified the correct model and reached a reasonable conclusion.

The Ten Equations offer a greater flexibility than the Ten Commandments. They can deal with a much larger range of problems and they give more nuanced advice. Am I putting the Ten Equations above the commandments of God? Yes, I am. Of course I am. We have had millennia to develop our thinking: we have come up with better ways of thinking about problems since the original Ten. I am putting the Ten Equations not only before Christianity, but also before many other approaches to life. The way data and model are combined, and nonsense is resisted, gives mathematics a raw honesty that allows it to rise above many other ways of thinking.

Mathematical knowledge is like an extra level of intelligence. I also believe – and this is more controversial – that it is our moral obligation to learn the Ten Equations. I even believe that, on the whole, the work the members of TEN has done so far is good for humanity. Not always, but more often than not. By learning the equations, you don't just help yourself, but you help others too.

This conclusion might come as a surprise, given that TEN's members so often have an edge over those who don't have the same skills. It may also appear to be a conclusion that is at odds with the philosophical position of verifiability described by A. J. Ayer, who told us that we can't expect to find sensible answers to moral questions in maths. But it is what I believe, and it is what I am going to argue for now. TEN is a force for good.

In order to discover where morality can be found in mathematics, first we have to be clear about where it can't be found. Through a process of elimination, we should be able to identify what it is about mathematical thinking that can tell us the 'right thing to do'.

Our last equation – 'if . . . then . . .' – is not a single equation, but shorthand for a set of algorithms that can be written using a series of 'if . . . then . . .' statements and 'repeat . . . until' loops. These statements are the basis of computer programming. Inside Anton's Soccer Bot, for example, we find commands like:

if *key passes* > *5* then *print ('He made a lot of important passes')*

Commands like this decide, in combination with input data, the output that is produced.

In the 1950s, 1960s and 1970s, the newly formed field of computer science discovered a whole range of algorithms for processing and organizing data. One of the earliest examples is the algorithm Merge sort, first proposed by John von Neumann in 1945 for sorting a list numerically or alphabetically. To understand how it works, first think about how to merge two already sorted lists. For example, I have one list containing {A,G,M,X} and another list with {C,E,H,V}. To create a sorted list that merges these two all I have to do is move left to right in both lists and place the letter that is first in the alphabet into a new list and then remove it from its original list.

Let's try this. First, I compare the first members of both lists, A and C. Since A comes first, I remove it from its current list and put it in a new to be sorted list. Now I have three lists: the new list {A} and the original lists {G,M,X} and {C,E,H,V}. Again I compare the first remaining members of the two original lists, G and C, and add C so that the new list becomes {A, C}. Next, I compare G and E, placing E in the new list to give {A, C, E}. This repeats until I have a sorted list: {A,C,E,G,H,M,V,X} and the original lists are empty.

To go from merging already sorted lists to sorting any lists, von Neumann proposed a strategy based on 'divide and conquer'. The full list is divided into smaller and smaller lists, which are each conquered using the same technique for merging already sorted lists.

Let's assume my original list is {X,G,A,M}. First we merge individual letters {X} and {G} to give {G,X}, and {A} and {M} to give {A,M}. Then the sorted list {A,G,M,X} is formed by merging {G,X} and {A,M}. The elegance of this approach is that it reuses the same technique on every level. By dividing our original list into sufficiently small parts, we eventually arrive at a list that is guaranteed to be sorted, i.e. individual letters. Then, by using the fact we already know how to merge two sorted lists, we guarantee that all the lists we create will also be sorted (Figure 10). Merge sort never gets it wrong.

Another example is Dijkstra's algorithm for finding the shortest path between two points. Edsger Dijkstra, a Dutch physicist and computer scientist, originally developed his algorithm in 1953 to demonstrate to 'non-computing people' (as he called them) that computers could be useful, by calculating the quickest driving route between two Dutch cities.[1] It took him only twenty minutes, sitting in a café in Amsterdam, to envision the algorithm. He later told the journal *Communications of the ACM* (Association for Computing Machinery) that, 'One of the reasons that it is so nice was that I designed it without pencil and paper. Without pencil and paper you are almost forced to avoid all avoidable complexities.'

Imagine you start in Rotterdam and want to drive to Groningen. Dijkstra's algorithm prescribes that you first label all Rotterdam's neighbouring towns with the travel time there from Rotterdam. Figure 10 illustrates this process. So, for example, Delft will take 23 minutes, Gouda, 28 minutes and Schoonhoven, 35 minutes. The next step is to look at all the neighbours of these three towns and find the shortest travel time to them. So, if it takes 35 minutes to get to Utrecht from Gouda and 32 minutes to get to Utrecht from Schoonhoven, then the shortest total travel time to Utrecht is $28 + 35 = 63$ minutes via Gouda (which is less than $35 + 32 = 67$ minutes via Schoonhoven). The algorithm keeps on extending outwards over the Netherlands, labelling the shortest distance to each town as it goes. Since the algorithm has calculated the shortest path to each town along the way, when a new town is added it is guaranteed that the shortest path to this new town will also be found. The algorithm doesn't set out to get to Groningen, it simply labels the distance to every town, but when it

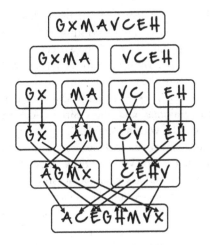

Mergesort is achieved by divide and conquer. First, the list of letters is divided into pairs, which are sorted. Sorted pairs are then merged into sorted lists of four, and so on until a completely sorted list is constructed.

In Dijkstra's shortest path algorithm, we move from one town to the next, finding the shortest path to each town along the way.

The black path is the shortest path between Rotterdam and Groningen. Numbers indicate travel time in minutes.

Figure 10: Illustration of Merge sort and Dijkstra's algorithm

finally adds Groningen into the calculation, then it is guaranteed to have found the shortest distance there.

The list of algorithms that are similar to von Neumann's Merge sort and Dijkstra's shortest path is very long.[2] Here are just a few examples: Kruskal's algorithm for finding a minimum spanning tree (the layout which uses the least track to connect all cities in a railway network); the Hamming distance for detecting differences between two pieces of text or data; the Convex Hull algorithm for drawing a shape around a collection of points; the Collision Detection algorithm for 3D graphics; or Fast Fourier Transforms for signal detection. These algorithms, and variations of them, are the building blocks of computer hardware and software. They sort and process our data, route our emails, check our grammar and allow Siri or Alexa to identify a song playing on the radio within seconds.

'If . . . then . . .' mathematics always gives the correct answer and we always know what it is going to do. Take the three Masters students' Soccer Bot, for example. I could ask it simple questions about football and it would answer, but for Anton the answers were not surprising. He has coded the rules that decide what the bot says and it reliably follows the rules he has laid out.

The reason I have lumped all these 'if . . . then . . .' algorithms together as a single equation is that they have one very important thing in common: they are universal truths. Dijkstra's algorithm will always find the shortest path; Merge sort will always sort a list of names from A to Z; and the Convex Hull of a set of points always has the same structure. They are statements that are true irrespective of what we say or do.

Throughout the first nine chapters of this book, we have used equations to test models, make predictions and sharpen our understanding of reality. These equations interact with the world: we allow past data to inform models and these models predict future data. By contrast, the 'if . . . then . . .' algorithms are inflexible recipes. They take in the data – say, a list of names to be sorted or a list of points between which the shortest path can be calculated – and they give an answer. We don't revise our knowledge of the world on the basis of the answers that are returned. Likewise, the truth of these algorithms is unaffected by our observations. That is why I call them universal: they have been proved to be true and they always work.

The examples I list above are the algorithms that underlie computer programming, but other mathematical theorems – about geometry, calculus and algebra – are also universally true. We saw one example in the friendship paradox in Chapter 5. At first it seemed inconceivable that our friends should be, on average, more popular than we are, but by reasoning logically about the question we showed that it was inevitable.

Mathematics is full of surprising results that may initially defy our intuition. For example, Euler's Identity (named after Leonhard Euler), $e^{\pi i} + 1 = 0$, tells us about a relationship between three well-known numbers, the exponential constant $e = 2.718...$, $\pi = 3.141...$, and $i = \sqrt{-1}$. The fact that it so elegantly combines such fundamental constants has led it to be called the most beautiful equation in maths.[3]

Another example is the Golden Ratio:

$$\phi = (1 + \sqrt{5})/2 = 1.618...$$

This number arises when we draw a rectangle which can be cut into a square and a new rectangle that is a scaled-down version of the original rectangle. Specifically, if the square has sides of length a and the rectangle has sides of length a and b, then a rectangle is golden if:

$$\frac{a+b}{a} = \frac{a}{b} = \phi$$

What is remarkable about ϕ is that it also appears in relationship to the Fibonacci sequence of numbers 1, 1, 2, 3, 5, 8, 13, 21, 34, ... obtained if we add the two previous numbers to get the next (i.e. $1 + 1 = 2$, $1 + 2 = 3$ and so on). If we take the ratio of two consecutive numbers, these get closer and closer to ϕ (i.e. $13/8 = 1.625$, $21/13 = 1.615...$, $34/21 = 1.619...$ and so on). These two examples are merely a starting point for a journey into pure mathematics, where much of our everyday intuition starts to fail us and rigorous logical reasoning is the only way forward.

The sheer number of mathematical theorems that have proved to be true led French mathematician Henri Poincaré to reflect in his 1902 book, *Science and Hypothesis*, that 'If all the assertions which mathematics puts forward can be derived from one another by formal logic, mathematics cannot amount to anything more than an immense tautology. Logical inference can teach us nothing essentially new . . . But can we really allow that these theorems which fill so many books serve no other purpose than to say in a roundabout fashion that "A = A"?' Poincaré's question was rhetorical, in that he believed that the challenges he and other humans faced in divining mathematical truths meant they must contain something deeper than just statements of logic.

A similar view underlies Dan Brown's *The Da Vinci Code*, the original but fictional account of a mathematical conspiracy theory. In the book, Professor Robert Langdon says, 'When the ancients discovered PHI [ϕ], they were certain they had stumbled across God's building block for the world . . . The mysterious magic inherent in the Divine Proportion was written at the beginning of time.' Langdon goes on to give examples (some of which are true and some of which aren't) of how the Golden Ratio, or Divine Proportion, as he calls it, has occurred in biology, art and culture. Throughout history, the members of TEN have used ϕ as a code, with the name of one of the novel's lead characters– SoPHIe Neveu – containing a clue.

I have to admit that I can find this aspect of mathematics enticing. I enjoyed *The Da Vinci Code* immensely. There is something incredible about the unexpected relationships we find, not only in numbers like phi, but also in Dijkstra's Shortest Path algorithm and von Neumann's Merge sort. There is a simple elegance that seems to transcend mundane reality. Could there be some deep code hidden within all these equations?

The correct answer to Poincaré's question turns out to be much more straightforward than he imagined. The answer is 'yes'. All the great theorems of mathematics, and the sorting and organizing algorithms of computer science, say nothing more than A does equal A. They are all just one immense tautology – very useful and unexpected tautologies, but tautologies none the less. Poincaré was literally correct and rhetorically wrong.

The argument that says Poincaré was 'correct' can be found in *Language, Truth and Logic*. In his book, Ayer used the example of a triangle. Imagine that a friend tells you about a triangle whose angles sum to less than 180 degrees.[4] You have two possible reactions: either to tell him he hasn't measured properly, or to tell him the object he is talking about isn't a triangle. Under no circumstances would you change your mind about the mathematical properties of a triangle on the basis of your friend's data. He won't find a triangle in the real world that undermines the results of geometry.

Similarly, a list of English words that can't be sorted alphabetically does not exist. If I show you a list with 'A. J. Ayer' placed after 'D. J. T. Sumpter' and tell you that it is the output of Merge sort, then you'll tell me that my algorithm isn't working or that I don't know the English alphabet. The list is certainly not evidence that Merge sort doesn't work. Nor is there a computer network through which the shortest path is longer than the second shortest path.

Unfortunately for Professor Langdon, the reason that all the different geometrical and mathematical relationships involve $\phi = 1.618...$ is that it is the positive root of the quadratic equation $x^2 - x - 1 = 0$. Solving the Fibonacci sequence and finding the Golden Ratio both involve solving this same quadratic equation and thus give rise to the same answer. There is no mysterious magic code hidden within phi or any other number.

Ayer's point is that mathematical theorems are independent of data. Maths is not verifiable. Instead, it consists of tautologies, statements that have been proven true by logic, but in themselves say nothing about reality. In response to the rhetorical tone used by Poincaré, Ayer wrote, 'The power of logic and mathematics to surprise us depends, like their usefulness, on the limitations of our reason.'

Poincaré was misled by the fact that it can be difficult to do mathematics, even for him. In fact, mathematical results are true independent of our observations. This is why I say that they are universal. They are true throughout the universe, independent of what we say and do, independent of scientific discoveries, independent of whether or not Poincaré or another mathematician has yet discovered them.

As we have seen throughout this book, the power of the Ten Equations lies in how they interact with the real world, by combining models and data. In isolation from data, equations do not have a deeper meaning. They certainly don't provide us with a morality or have anything to do with God. They are just a bunch of extremely useful results that happen to be true.

To find mystery and morality in maths we are going to have to look in places other than the theories themselves.

I'd been putting off calling Marius. Jan had asked me to check with him before I printed the financial details of their betting operation and I was a bit scared Marius might say 'no'. He might want to keep the secrets safe from the public gaze.

I needn't have worried. Marius was happy to talk and filled me in on the details. Their profits were still growing by the day, though it had been tough for him. 'If you look at the numbers day by day you go crazy,' he said. At one point they had had a bad run, losing $40,000. 'It's the worst experience ever. You really start to get doubts. But we just kept confidence testing and let time go by. And the numbers came back up again. Then down again . . . then up again.'

He told me that gambling had taught him to be more patient, to focus on the things he could change. 'We can't control the fluctuations. I've learnt not to watch the games like we did during the World Cup. At first, I'd check our bets live. Now, I just come into the office and work and we review quarterly.'

'Do you ever think of the morality of what you are doing?' I asked him. 'All those people who are losing while you win?'

'I think that gambling is tricky because of all the misleading adverts,' Marius said. 'But at the same time, just a quick search online about value betting and you can find all you need to know to remain profitable as an amateur. People just aren't willing to put in the effort.'

He was right. This is the moral lesson from the betting equation. If people aren't willing to spend a few hours searching out information on the Internet, then why should it be Marius's responsibility? Marius and Jan had created a website containing exactly the information these people needed to make value bets at the soft bookmakers. Yet very few people bet in the way they do.

I asked Marius what he would do if the markets changed unexpectedly and he lost everything. 'You can never know how things will turn out. That's always a possibility,' he said. 'But I very much enjoy what I do and that is where happiness comes from. I would never be satisfied sitting on a beach while a bot does the work. What I find exciting is digging through the data and finding things out.' He had found out the real secret of TEN and it had little to do with how much money he had in his bank account. His reward was in how much he had learnt.

Is this a sign of morality? I think it is. I believe there is an intellectual honesty in Jan and Marius's approach. They don't lie about what they are doing. They play the game as it is set up and they win because they are better at it. The same reasoning applies to William Benter and Matthew Benham's betting operations on a larger scale. Benter's honesty was astounding. Admittedly, he was cautious about revealing how much he'd won and where, but he published his methods in a scientific journal. Anyone else with the mathematical skills can now pick up Benter's methods and apply them.

Those who work hard, who learn and who persevere are the winners. Those who cut corners lose. This same rule applies throughout TEN. When we are making judgements, we are forced to say how our beliefs have been shaped by data. When we are building a model of skill, we are forced to state our assumptions. When we make an investment or place a bet, we are forced to admit our profits and our losses in order to improve our model. We are forced to tell each other how confident we are about our conclusions. We are forced to admit we are not at the centre of our social network, and we shouldn't feel sorry for ourselves about being less popular. When we see correlation, we are forced to search for causation. When we build technology, we are shown how it rewards and punishes the people who use it. This is the hard morality of mathematics. The truth always wins, eventually.

The members of TEN are the true guardians of intellectual honesty. They set out their assumptions, they collect the data and they tell us the answers. When they don't know the full answer, they tell us what is missing. They list the plausible alternatives and the probability of success of each of them, and they start thinking about the next steps we can take to find out more.

Put honesty back into your own life. The Ten Equations can help you. It starts by thinking about the probabilities, both by gambling to get what you want and by understanding the risk of failure. Improve your judgement by collecting data before you draw conclusions. Improve your confidence, not by convincing yourself that you are right, but by spinning the wheel many times. Each lesson the equations teach us, from revealing the filter created by our social network to understanding how social media drives us to a tipping point, reiterates the importance of being honest about our model and ensuring we use data to improve ourselves.

If you follow these equations, you will notice how those around you will come to respect your judgement and your patience. This is the first sense in which mathematics can be a source of morality. It is the deliverer of hard truths about yourself and those around you.

*

Ben Rogers' biography of A. J. Ayer recounts a story about the meeting in 1987 between the philosopher and the boxer Mike Tyson.[5] Ayer, then aged seventy-seven, was at a party on West 57th Street in Manhattan when a woman ran into the room saying that her friend was being assaulted in a bedroom in the apartment. When Ayer investigated, he found Tyson trying to force himself on a young, soon-to-be supermodel, Naomi Campbell.

Rogers writes that Ayer warned Tyson to desist. To which Tyson replied, 'Do you know who the fuck I am? I'm the heavyweight champion of the world.'

Ayer stood his ground. 'And I am the former Wykeham Professor of Logic. We are both pre-eminent in our field. I suggest that we talk about this like rational men.'

Tyson, who was apparently a fan of philosophy, was duly impressed and backed down.

If Tyson had wanted to hit Ayer with an intellectual uppercut, though, he could have asked him on what basis the philosopher felt it was justified to interfere with his advances to Naomi Campbell. After all, as Ayer had argued in *Language, Truth and Logic*, morality lies outside the scope of an empirical discussion. While it might well have been the case that Campbell was scared of Tyson, Iron Mike could have asked, 'Is there any logical reason why it isn't OK for men to force themselves on women in their own pursuit of sexual desire?'

Ayer would have had to concede that he was simply imposing the agreed-upon norms of the type of social gathering at which they were currently present. To which Tyson could append that his norms, acquired through an early life of petty crime on the streets of Brooklyn, were different from those of Eton-educated Ayer, thus giving them little common ground upon which to continue their discussion. 'And if you don't mind,' Iron Mike might have continued, 'I would like to get back to wooing this beautiful lady in the way I think most appropriate.'

Whether this was the direction of their conversation, I can't possibly know. What is known is that the two of them fell into a discussion, during which Naomi Campbell escaped from the party. Mike Tyson was convicted for the rape of another woman four years later and is now a registered sex offender.

The Tyson v. Ayer story illustrates (among other things) the fundamental problem for anyone following the strictly logical positivist approach. It becomes impossible to solve even the most obvious of moral dilemmas. For all the honesty of mathematics and logical thought, every individual is left alone to decide their own morality.

Clearly something is missing from logical positivism here. The question is what? In order to sharpen our thinking about the role of maths in morality, in 1967 Philippa Foot, an English moral philosopher from Oxford University, developed a thought experiment that has become known as the trolley problem.[6] It can be described as follows:

> Edward is the driver of a trolley whose brakes have just failed. On the track ahead of him are five people: the banks are so steep that they will not be able to get off the track in time. The track has a spur leading off to the right, and Edward can turn the trolley onto it. Unfortunately, there is one person on the right-hand track. Edward can turn the trolley, killing the one; or he can refrain from turning the trolley, killing the five.

The question is what Edward should do: divert the trolley and kill one or continue straight ahead and kill five. After some thought, most of us tend to agree that he should go for the first option. It is better to kill one person than five. So far, so good.

Now consider another trolley problem posed by Judith Thomson, Professor of Philosophy at MIT, in 1976:

George is on a footbridge over the trolley tracks. He knows trolleys, and can see that the one approaching the bridge is out of control. On the track back of the bridge there are five people; the banks are so steep that they will not be able to get off the track in time. George knows that the only way to stop an out-of-control trolley is to drop a very heavy weight into its path. But the only available, sufficiently heavy weight is a fat man, also watching the trolley from the footbridge. George can shove the fat man on to the track in the path of the trolley, killing the fat man; or he can refrain from doing this, letting the five die.

What should George do? On the one hand, it is obviously wrong to shove a man on to a railway track. On the other hand, by failing to act, he is allowing five to die, just as Edward would have done if he hadn't diverted the trolley.

In a survey, roughly 81% of 1,000 American citizens said that if they were Edward then they would divert the trolley and kill one person, while only 39% thought that George should push the man and save five people.[7] Chinese and Russia respondents to the same question were also more inclined to think that Edward should act and George should not, supporting the hypothesis that we share a universal moral intuition about these dilemmas.[8] Cultural differences do remain, however, with Chinese people more likely to let the trolley take its course in both cases.

Judith Thomson developed the second trolley problem in order to make the dilemma crystal clear.[9] Both problems describe exactly the same mathematical problem, saving five lives against saving one life, but our intuition tells us that they are very different from each other. The mathematical solution is simple, but the moral solution is more complex. The trolley problems ask us to think about the actions we are prepared – and not prepared – to carry out to save lives.

Trolley problems lie at the centre of much of modern science fiction. The dilemma is usually revealed about thirty minutes to an hour into the film (spoiler alert). In the Marvel Comics movie, *Avengers: Infinity War*, philosophically minded super-baddy Thanos decides, after witnessing how over-population has completely destroyed his home planet, that it would be a good idea to kill half the Universe's population. He reasons that killing half now will save more later and in the terms of Thomson's trolley dilemma decides to throw billions of fat men under a trolley with a single click of the finger. The sequel, *Avengers: Endgame*, sees Tony Stark face a more personal dilemma of the same type: his daughter's existence is now held in the balance versus bringing his friends back. He is asked to make an almost impossible choice between the two.

Usually in science fiction it is the baddies who choose to push the fat guy. In many cases the decision is portrayed as a form of brutal, merciless logic. Robots or rogue AIs make the utilitarian decision to save five, rather than one, no matter how horrible the action that must be carried out in order to achieve their goal. For a utilitarian robot, which humans have programmed to save as many lives as possible, it is numbers and not feelings which are prioritized. All else being equal, the utility of five people is greater than the utility of one.

It is exactly the same issue that Philippa Foot and Judith Thomson illustrated with the trolley problem: it is wrong to believe that we can solve such dilemmas using a utilitarian approach. The robots are wrong in the films and, if they existed, they would be wrong in real life too. Science fiction reminds us that we can't create a universal rule, such as

if $5 > 1$ then *print('save the 5')*

to solve all of life's problems. If we do so we'll end up performing the most terrible moral crimes, which we would never be able to justify to future generations.

When I was younger, I might have viewed a failure to act – even if the act involved pushing a fat man on to a railway track – as an indication of logical weakness on the part of humanity.

That would have been wrong of me, and not simply because I was being too hard on my fellow humans. In fact, it would have been a logical weakness on my part to draw that conclusion. The trolley problem tells us two things. First, it reinforces the fact that there are no purely mathematical answers to questions about real life. This was the same point that Ayer made in reply to Poincaré's rhetorical question about the 'universal' nature of mathematics. It is the reason that the Da Vinci code doesn't exist. Our feeling of universality about maths is a result of its tautological nature, not a deeper truth. We cannot use maths as divine commandments. We can only use it, as we have in this book, as a tool for organizing our work around models and data.

The second thing that the trolley problem tells us is that pure utilitarianism – the idea that our morals should be built around an attempt to maximize human life or happiness or any other variable – is one of the greater evils among all the (equally incorrect) moral codes available to us.[10] A rule such as 'save as many human lives as possible' very quickly contradicts our moral intuition and causes us to do horrible things. As soon as we start setting out optimal moral codes, we end up creating a moral maze.

I have come to believe that there is a very simple answer to these dilemmas: we should learn to trust and use our own moral intuitions. This is what A. J. Ayer did when he confronted Mike Tyson. It is what Moa Bursell did when she decided to investigate racism after seeing her friends chased by Nazis. It is what has guided me when I looked into Cambridge Analytica, fake news and algorithmic bias; or when I supervised Björn's thesis on immigration and the rise of the right in Sweden. It is what guided Nicole Nisbett when she looked at political communication. It is also what Spider-Man does: he follows his intuition and uses his skills to kill the bad guy.

The trolley problem tells us that we need a softer way of thinking about such moral and philosophical dilemmas, a way that complements the hard, brutal honesty of applying models and data.

The members of TEN who contribute most to society think both soft – using their moral intuition to decide which problems to solve; and hard – combining models and data to be honest in their answers. They have listened and understood the values of those around them. They realize that they are not any more qualified than anyone else to decide which problems need to be solved, but they are more qualified to solve them. They are public servants who have kept the spirit that Richard Price introduced to TEN almost 260 years ago. Price was wrong about miracles,[11] but he was right about the need for morality in the way we apply maths.

I don't have any conclusive proof, but I believe that after logical positivism strips away the idea of a universal utilitarianism, we are left with moral intuition to guide us. It is this soft way of thinking that tells us which problems to solve.

*

The members of TEN need to talk. We need to learn how to handle the power that has been handed down to us – in the same way that Spider-Man realizes his weaknesses during each new incarnation.

Being 'soft' means it is not OK to mindlessly multiply the money of ignorant investment bankers. It means we shouldn't patent basic equations in order to make profits. It means that we should continue to be open about the algorithms we use, that we should share all our secrets with those who are prepared to put in the hard work to learn them.

We have to use our intuition to lead us towards the questions that are important. We should listen to the feelings of others, and find out what is important to them. Many of us do this already, but we need to be open about what we are and why we do what we do.

We need to be soft when we define our problems and brutally hard when we solve them.

*

I am sitting in a seminar room in the basement of a maths tower in a university in the north of England. Viktoria Spaiser, academic fellow in Politics at the University of Leeds, is standing in front of us, introducing the day's speakers. She, and her partner in research and in life, Richard Mann, have organized a two-day workshop on maths for social activism. The idea is to bring together mathematicians, data scientists, public policy-makers and people from business to find ways to use mathematical models to make the world a better place.

I first met Viktoria almost eight years ago now; Richard a little while before that. Alongside one of my PhD students, Shyam Ranganathan, we have worked together on modelling racial segregation in schools in Sweden,[12] democratic change across nations, and on working out how the United Nations can reach its often-contradictory sustainable development goals. We haven't always said this out loud, but we have always secretly believed that maths shouldn't just study the world: it should change it for the better. The choice of the word 'activism' in the title of this meeting is our first attempt to be open about our goals.

We are not alone. After Viktoria opens the meeting, the participants stand up, one after another, and tell us what they are up to. Adam Hill from DataKind, UK, has created a network showing links between the people sitting on the boards of anonymous companies, set up to conceal ownership, in the UK. He and his team can detect corruption and potential money laundering by coupling together ownerships. Betty Nannyonga tells us how her colleagues are using mathematical models to understand the causes of student strikes at Makerere University, Kampala, where she works in Uganda. Anne Owen, academic fellow at the University of Leeds, has shown that Greta Thunberg was correct when she claimed that the UK has been dishonest about claims for its reduction in CO_2 emissions.[13] Anne shows us the proper calculation, which accounts for the production and transport of all the plastic goods we import from China. The average person aged sixty to sixty-nine is responsible – by flying off on holiday or driving a large car – for releasing 64% more CO_2 per year than the average young person under thirty. It is this older generation, some of them who are criticizing Thunberg, who need to think most carefully about their carbon footprint.

You might not have known about us until now, but now you do. The secret is out.

We are TEN.

Notes

I THE BETTING EQUATION

1   The article in question was published on the publishing platform Medium: see <https://medium.com/@Soccermatics/if-you-had-followed-the-betting-advice-in-soccermatics-you-would-now-be-very-rich-1f643a4f5a23>. A full description of the model can be found in my book *Soccermatics: Mathematical Adventures in the Beautiful Game* (London: Bloomsbury Publishing, 2016).

2   Each bet multiplies your capital by 1.0003 (0.03% increase per bet). If you make 100 bets a day throughout the year, then your expected capital at the end of the year is $1,000 \times 1.0003^{100 \times 365} = 56,860,593.80$.

3   A bookmaker's odds are fair if the odds *for* an event taking place multiplied by the odds *against* the event occurring is equal to 1. For example, if the odds are 3/2 then the odds for a draw and a victory by the underdog must sum to 2/3, because $3/2 \times 2/3 = 1$. In practice, bookmakers never offer fair odds. So, in the above example, they will tend to offer 7/5 for the favourite to win and 4/7 against the favourite winning, so that $7/5 \times 4/7 < 1$. The bookmaker's margin in this case is $1/(1 + 7/5) + 1/(1 + 4/7) - 1 = 0.05$ or 5%.

4   Your expected profit per bet is

$$\frac{2}{5} \times \frac{7}{5} + \frac{3}{5} \times -1 = \frac{14}{25} - \frac{15}{25} = -\frac{1}{25}$$

i. e. 4 cents are lost per bet.

5   Even after 5 failed attempts, you shouldn't be too despondent. If each interview has a 1 in 5 success probability, then the probability it takes at least 5 interviews until you succeed is $(1 - 1/5)^5 = 33\%$.

6   William Benter, 'Computer based horse race handicapping and wagering systems: a report', in Donald B. Hausch, Victor S. Y. Lo and

William T. Ziemba (eds), *Efficiency of Racetrack Betting Markets* revised edn (Singapore: World Scientific Publishing Co. Pte Ltd, 2008), pp. 183–98.

7   Kit Chellel, 'The gambler who cracked the horse-racing code', *Bloomberg Businessweek*, 3 May 2018; at <https://www.bloomberg.com/news/features/2018-05-03/the-gambler-who-cracked-the-horse-racing-code>.

8   Ruth N. Bolton and Randall G. Chapman, 'Searching for positive returns at the track: a multinomial logit model for handicapping horse races', *Management Science* 32(8) (August 1986): 1040–60.

9   David R. Cox, 'The regression analysis of binary sequences', *Journal of the Royal Statistical Society: Series B (Methodological)* 20(2) (1958): 215–32.

2 THE JUDGEMENT EQUATION

1   One in 10 million shouldn't be taken as a precise value. The Civil Aviation Authority report *Global Fatal Accident Review 2002 to 2011*, CAP 1036 (June 2013), estimated 0.6 fatal accidents per million flights flown, excluding terrorist attacks, per flight flown during the period 2002 to 2011. Not everyone dies in a fatal accident and the statistics vary between countries, so it is difficult to give an exact number. It is, in any case, small on a scale of millionths.

2   The derivation of this equation from Bayes' rule (Equation 2) requires an understanding of integrals. For a measurement like θ, that can take continuous values between 0 and 1, Bayes' rule is written:

$$p(\theta|D) = \frac{P(D|\theta) \cdot p(\theta)}{\int_0^1 P(D|x) \cdot p(x)dx}$$

where the function $p( )$ is known as a density function. The integral on the bottom is over all possible values of θ and plays the same role as the sum in Equation 2. From this we can write:

$$P(\theta > 0.99 \,|\, 100 \text{ sunrises})$$

$$= \int_{0.99}^1 p(\theta|100 \text{ sunrises})d\theta = \frac{\int_{0.99}^1 p(100 \text{ sunrises} \,|\, \theta) \cdot p(\theta)d\theta}{\int_0^1 p(100 \text{ sunrises} \,|\, x) \cdot p(x)dx}$$

From here we can note that $p(100 \text{ sunrises} \mid \theta) = \theta^{100}$ is the probability we get 100 sunrises in a row if the probability of a sunrise on a particular day is θ. We then set the term $p(x) = 1$. This says that, before the man arrives on Earth, all values of x are equally likely, an assumption set out in the way Bayes describes the problem of the newly arrived man. Putting these values into the equation above gives:

$$p(\theta > 0.99 \mid 100 \text{ sunrises}) = \frac{\int_{0.99}^{1} \theta^{100} \cdot 1 d\theta}{\int_{0}^{1} \theta^{100} \cdot 1 dx} = \frac{(1 - 0.99^{101})/101}{1/101} = 1 - 0.99^{101} \approx 0.638$$

as in the text.

3 This result is highly counter-intuitive, but it is mathematically correct. To further convince yourself, assume that $\theta = 0.98$ and the true probability of a sunrise is 98%. If this were the case, it wouldn't be that surprising if the sun rose on all of the 100 days he observed. The probability of 100 sunrises in a row in this case is $0.98^{100} = 13.3\%$. Small but not negligible. The same logic applies to $\theta = 0.985$ $(0.985^{100} = 22.1\%)$ and for other values of $\theta < 0.99$. Although the value of θ is more likely to be greater than 99%, it remains reasonably likely (with the probability of 36.2% to be exact) that it is smaller than 99%.

4 David Hume, *An Enquiry Concerning Human Understanding* (London, 1748).

5 This quote, and the argument in this paragraph, is adapted from David Owen, 'Hume *versus* Price on miracles and prior probabilities: testimony and the Bayesian calculation', *Philosophical Quarterly* 37(147) (April 1987): 187–202.

6 This calculation is left to the interested reader. Use note 2 above to help you.

7 As above. Use note 2 to help you.

8 Martha K. Zebrowski, 'Richard Price: British Platonist of the eighteenth century', *Journal of the History of Ideas* 55(1) (January 1994): 17–35.

9 Richard Price, *Observations on Reversionary Payments . . . To Which Are Added, Four Essays on Different Subjects in the Doctrine of Life-Annuities . . . A New Edition, With a Supplement, etc.*, Vol. 2 (London: T. Cadell, 1792).

10 Geoffrey Poitras, 'Richard Price, miracles and the origins of Bayesian decision theory', *European Journal of the History of Economic Thought* 20(1) (February 2013): 29–57.

11 Richard Price and Anne-Robert-Jacques Turgot, *Observations on the Importance of the American Revolution, and the Means of Making it a Benefit to the World* (London: T. Cadell, 1785).

12 Ian Vernon, Michael Goldstein and Richard G. Bower, 'Galaxy formation: a Bayesian uncertainty analysis', *Bayesian Analysis* 5(4) (2010): 619–69.

13 Christine Carter, 'Is screen time toxic for teenagers?', *Greater Good Magazine*, 27 August 2018; at <https://greatergood.berkeley.edu/article/item/is_screen_time_toxic_for_teenagers>.

14 Candice L. Odgers, 'Smartphones are bad for some adolescents, not all', *Nature* 554(7693) (February 2018): 432–4.

15 This result is originally from a study of UK teenagers; see Andrew K. Przybylski and Netta Weinstein, 'A large-scale test of the Goldilocks hypothesis: quantifying the relations between digital-screen use and the mental well-being of adolescents', *Psychological Science* 28(2) (January 2017): 204–15.

3 THE CONFIDENCE EQUATION

1 The equation for the Normal curve, which de Moivre wrote down (in logarithmic form) in the second edition of his book on probability in 1738 is:

$$\frac{1}{\sqrt{2\pi\sigma^2}} \, \exp\left(-\frac{(x-\mu)^2}{2\sigma^2}\right)$$

where μ is the mean and σ is the standard deviation.

2 The third and final version is available on Google Books. Abraham de Moivre, *The Doctrine of Chances: Or, A Method of Calculating the Probabilities of Events in Play. The Third Edition* (London: A. Millar, 1756).

3 The probability of getting two aces in a five-card hand involves first multiplying the probability of being dealt an ace with the first card (4/52) with the probability of also being dealt an ace on the second card (3/51), followed by the probability of getting no aces for the next three cards which are, respectively, 48/50, 47/49 and 46/48. This gives us the probability of getting dealt the two aces first and then the other three non-aces, but we also need to remember that there are ten different potential orderings of the 2 aces in the 5-card hand. Thus, the overall probability is:

$$10 \cdot \frac{4 \cdot 3 \cdot 48 \cdot 47 \cdot 46}{52 \cdot 51 \cdot 50 \cdot 49 \cdot 48} = \frac{259440}{6497400} = \frac{2162}{54145} = 4\%$$

4 .Helen M. Walker, 'De Moivre on the law of normal probability' (2006); at <https://www.semanticscholar.org/paper/DE-MOIVRE-ON-THE-LAW-OF-NORMAL-PROBABILITY-Walker/d40c10d50e86f0ceed1a059d81080a3bd9b56ffd#citing-papers>.

5 The history of the CLT is reviewed in Lucien Le Cam, 'The central limit theorem around 1935', *Statistical Science* 1(1) (1986): 78–91.

6 There is a caveat here. Each measurement has to have a finite mean and standard deviation for the result to apply.

7 Statistics sourced from <https://stats.nba.com/search/team-game/>.

8 Richard E. Just and Quinn Weninger, 'Are crop yields normally distributed?' *American Journal of Agricultural Economics* 81(2) (May 1999): 287–304.

9 Nate Silver, *The Signal and the Noise: The Art and Science of Prediction* (London: Allen Lane, 2012).

10 $\sigma^2 = \frac{1}{3} \cdot (0 - (-1))^2 + \frac{2}{3} \cdot (0 - \frac{1}{2})^2 = \frac{1}{3} + \frac{1}{6} = \frac{1}{2}.$

So the standard deviation is $\sigma = 0.71$.

11 These values provide the number of observations to allow us to be 97.5% (rather than 95%) certain that we haven't made a mistake and h is actually 0 or less. The 97.5% certainty arises because our 95% confidence interval covers both lower and upper limits of h. There is also a 2.5% chance that we have underestimated our edge, and that our edge is bigger than our confidence interval suggests. But underestimating our edge isn't a problem as far as profitable gambling is concerned, so it is only the 2.5% chance that we have overestimated our edge that is a concern. They also assume the edge is positive, $h > 0$. But the same result applies for a negative edge, with $-h$ in place of h.

12 I checked the standard deviation for several hotels and found that it is often slightly less than 1 – about 0.8, for example. But assuming it is 1 is reasonable enough.

13 Mahmood Arai, Moa Bursell and Lena Nekby, 'The reverse gender gap in ethnic discrimination: employer stereotypes of men and women with Arabic names', *International Migration Review* 50(2) (2016): 385–412.

14 The variance in the response to foreign born names is

$$\sigma_F{}^2 = \frac{43}{187}\left(1 - \frac{43}{187}\right)^2 + \frac{(187 - 43)}{187}\left(0 - \frac{43}{187}\right)^2 = 0.177$$

while the variance in response to Swedish names is

$$\sigma_S^2 = \frac{79}{187}\left(1 - \frac{79}{187}\right)^2 + \frac{(187 - 79)}{187}\left(0 - \frac{79}{187}\right)^2 = 0.244$$

The total variance is thus $\sigma^2 = \sigma_F^2 + \sigma_S^2 = 0.177 + 0.244 = 0.421$ and thus $\sigma = 0.6488$. Special thanks to Rolf Larsson for spotting an earlier mistake I made in this calculation.

15 Marianne Bertrand and Sendhil Mullainathan, 'Are Emily and Greg more employable than Lakisha and Jamal? A field experiment on labor market discrimination', *American Economic Review* 94(4) (September 2004): 991–1013.

16 Zinzi D. Bailey, Nancy Krieger, Madina Agénor, Jasmine Graves, Natalia Linos and Mary T. Bassett, 'Structural racism and health inequities in the USA: evidence and interventions', *The Lancet* 389(10077) (April 2017): 1453–63.

17 This is a rule of thumb that I find useful but which requires some mathematical justification. In the example, the population as a whole has proportion p of a certain type (say white people). The variance is maximized when $p = 1/2$, so for all values of p the variance is less than $1/2(1 - 1/2) = 1/4$ and thus the standard deviation is less than $1/2$. Since 1.96 is almost equal to 2, this means that the confidence interval for the sample proportion p^* is $1.96\frac{1/2}{\sqrt{n}} \approx 1/\sqrt{n}$. Hence the rule of thumb.

18 This is discussed in, for example, Karl Pearson, 'Historical note on the origin of the Normal Curve of Errors', *Biometrika* 16(3–4) (December 1924): 402–4.

19 Tukufu Zuberi and Eduardo Bonilla-Silva (eds), *White Logic, White Methods: Racism and Methodology* (Lanham, MD: Rowman & Littlefield Publishers, 2008).

20 John Staddon, 'The devolution of social science', *Quillette*, 7 October 2018; at <https://quillette.com/2018/10/07/the-devolution-of-social-science/>.

21 Jordan B. Peterson, *12 Rules for Life: An Antidote to Chaos* (Toronto, ON: Penguin Random House Canada, 2018).

22 For example in an interview on Scandinavian TV chat show *Skavlan* in November 2018.

23 Katrin Auspurg, Thomas Hinz and Carsten Sauer, 'Why should women get less? Evidence on the gender pay gap from multifactorial survey experiments', *American Sociological Review* 82(1) (2017): 179–210.

24 Corinne A. Moss-Racusin, John F. Dovidio, Victoria L. Brescoll, Mark J. Graham and Jo Handelsman, 'Science faculty's subtle gender biases favor male students', *Proceedings of the National Academy of Sciences* 109(41) (October 2012): 16474–9.

25 Eric P. Bettinger and Bridget Terry Long, 'Do faculty serve as role models? The impact of instructor gender on female students', *American Economic Review* 95(2) (May 2005): 152–7.

26 Allison Master, Sapna Cheryan and Andrew N. Meltzoff, 'Computing whether she belongs: stereotypes undermine girls' interest and sense of belonging in computer science', *Journal of Educational Psychology* 108(3) (April 2016): 424–37.

27 John A. Ross, Garth Scott and Catherine D. Bruce, 'The gender confidence gap in fractions knowledge: gender differences in student belief–achievement relationships', *School Science and Mathematics* 112(5) (May 2012): 278–88.

28 Emily T. Amanatullah and Michael W. Morris, 'Negotiating gender roles: gender differences in assertive negotiating are mediated by women's fear of backlash and attenuated when negotiating on behalf of others', *Journal of Personality and Social Psychology* 98(2) (February 2010): 256–67.

29 For a comprehensive review of these issues in maths and engineering, read both Sapna Cheryan, Sianna A. Ziegler, Amanda K. Montoya and Lily Jiang, 'Why are some STEM fields more gender balanced than others?', *Psychological Bulletin* 143(1) (January 2017): 1–35; and Stephen J. Ceci, Donna K. Ginther, Shulamit Kahn and Wendy M. Williams, 'Women in academic science: a changing landscape', *Psychological Science in the Public Interest* 15(3) (November 2014): 75–141.

30 Transcript taken from Conor Friedersdorf, 'Why can't people hear what Jordan Peterson is saying?', *The Atlantic*, 22 January 2018; at <https://www.theatlantic.com/politics/archive/2018/01/putting-monsterpaint-onjordan-peterson/550859/>.

31 A good academic introduction to this methodology is Peter Hedström and Peter Bearman (eds), *The Oxford Handbook of Analytical Sociology* (Oxford: Oxford University Press, 2011).

32 Joseph C. Rode, Marne L. Arthaud-Day, Christine H. Mooney, Janet P. Near and Timothy T. Baldwin, 'Ability and personality predictors of salary, perceived job success, and perceived career success in the initial career stage', *International Journal of Selection and Assessment* 16(3) (September 2008): 292–9.

33 If you do pedantically insist on using the 63% to 37% information, that's fine, but you then have to be consistent in your approach. You need to hop back one chapter in this book and apply the judgement equation when you talk to Jane and Jack. You should start by setting your model M to be 'Jane is more agreeable than Jack' with $P(M) = 63\%$. Now walk into the room, smile and talk to both of them in the same way. Even a small amount of eye contact and a few sentences will give

you a fair amount of data *D* about their agreeableness. Now you can update $P(M|D)$ and make a better judgement. The original $P(M)$ will quickly become irrelevant.

34 In an interview on Scandinavian TV chat show *Skavlan* in November 2018.

35 These quotes are from a blog post by Peterson, written in February 2019: 'The gender scandal: part one (Scandinavia) and part two (Canada)'; at <https://www.jordanbpeterson.com/political-correctness/the-gender-scandal-part-one-scandinavia-and-part-two-canada/>.

36 Janet Shibley Hyde, 'The gender similarities hypothesis', *American Psychologist* 60(6) (September 2005): 581–92.

37 Ethan Zell, Zlatan Krizan and Sabrina R. Teeter, 'Evaluating gender similarities and differences using metasynthesis', *American Psychologist* 70(1) (January 2015): 10–20.

38 Janet Shibley Hyde, 'Gender similarities and differences', *Annual Review of Psychology* 65 (January 2014): 373–98.

39 Gina Rippon, *The Gendered Brain: The New Neuroscience that Shatters the Myth of the Female Brain* (London: Bodley Head, 2019).

4 THE SKILL EQUATION

1 You can watch Ayer tell the story himself here: 'A. J. Ayer on Logical Positivism and its legacy' (1976); at <https://www.youtube.com/watch?v=nGoEWNezFl4>.

2 Kevin Reichard, 'Measuring MLB's winners and losers in costs per win', *Ballpark Digest*, 8 October 2013; at <https://ballparkdigest.com/201310086690/major-league-baseball/news/measuring-mlbs-winner-and-losers-in-costs-per-win>.

3 George R. Lindsey, 'An investigation of strategies in baseball', *Operations Research* 11(4) (July–August 1963): 477–501.

4 Bruce Schoenfeld, 'How data (and some breathtaking soccer) brought Liverpool to the cusp of glory', *New York Times Magazine*, 22 May 2019; at <https://www.nytimes.com/2019/05/22/magazine/soccer-data-liverpool.html>.

5 THE INFLUENCER EQUATION

1 I use the United Nations definitions of city sizes here; see *The World's Cities in 2018—Data Booklet* (ST/ESA/SER.A/417), United Nations,

Department of Economic and Social Affairs, Population Division (2018).

2   The matrix is multiplied as follows:

$$\begin{pmatrix} 0 & 1/2 & 0 & 0 & 0 \\ 1/2 & 0 & 1/3 & 1/3 & 1/3 \\ 1/2 & 1/2 & 0 & 1/3 & 1/3 \\ 0 & 0 & 1/3 & 0 & 1/3 \\ 0 & 0 & 1/3 & 1/3 & 0 \end{pmatrix} \begin{pmatrix} 1 \\ 0 \\ 0 \\ 0 \\ 0 \end{pmatrix} = \begin{pmatrix} 0 \cdot 1 + 1/2 \cdot 0 + 0 \cdot 0 + 0 \cdot 0 + 0 \cdot 0 \\ 1/2 \cdot 1 + 0 \cdot 0 + 1/3 \cdot 0 + 1/3 \cdot 0 + 1/3 \cdot 0 \\ 1/2 \cdot 1 + 1/2 \cdot 0 + 0 \cdot 0 + 1/3 \cdot 0 + 1/3 \cdot 0 \\ 0 \cdot 1 + 0 \cdot 0 + 1/3 \cdot 0 + 0 \cdot 0 + 1/3 \cdot 0 \\ 0 \cdot 1 + 0 \cdot 0 + 1/3 \cdot 0 + 1/3 \cdot 0 + 0 \cdot 0 \end{pmatrix} = \begin{pmatrix} 0 \\ 1/2 \\ 1/2 \\ 0 \\ 0 \end{pmatrix}$$

The other matrix multiplications are performed using the same rule. Each item in each row in the matrix is multiplied by the items in the column vector, and the new vector is the sum of the item-wise multiplication.

3   To learn more about this research area from an academic perspective, I recommend Mark Newman, *Networks*, 2nd edition (Oxford: Oxford University Press, 2018).

4   Scott L. Feld, 'Why your friends have more friends than you do', *American Journal of Sociology* 96(6) (1991): 1464–77.

5   Here I show this result more rigorously. Let $P(X_i = k)$ be the probability that individual i has k followers. Now, consider first choosing one individual j and then choosing i from the people that j follows. The probability that the person i has X_i followers can be written as $P(X_i = k)$. We can calculate this probability using Bayes' theorem (Equation 2):

$$P(X_i = k \mid j \text{ follows } i) = \frac{P(j \text{ follows } i \mid X_i = k) \cdot P(X_i = k)}{\Sigma_{k'} P(j \text{ follows } i \mid X_i = k') \cdot P(X_i = k')}$$

We know that $P(j \text{ follows } i \mid X_i = k) = k/N$, where N is the total number of edges in the graph. Therefore:

$$P(X_i = k \mid j \text{ follows } i) = \frac{(k/N) \cdot P(X_i = k)}{\Sigma_{k'}(k'/N) \cdot P(X_i = k')} = \frac{k \cdot P(X_i = k)}{\Sigma_{k'} k' \cdot P(X_i = k')} = \frac{k \cdot P(X_i = k)}{E[X_i]}$$

Thus when $k > E[X_i]$ then $P(X_i = k \mid j \text{ follows } i) > P(X_i = k)$ and, likewise, when $k < E[X_i]$ then $P(X_i = k \mid j \text{ follows } i) < P(X_i = k)$. This tells us that a randomly chosen person followed by another randomly chosen person is likely to have more followers than a person chosen at random.

To show that a randomly chosen person has fewer followers than the average person they follow, we calculate the expected (average) number of followers of all of the people followed by j. This is given by:

$$E[X_i = k \mid j \text{ follows } i] = \Sigma_k k \cdot P(X_i = k \mid j \text{ follows } i) = \Sigma_k \frac{k^2 \cdot P(X_i = k)}{E[X_i]} = \Sigma_k \frac{E[X_i]^2 + \text{Var}[X_i]}{E[X_i]}$$

Thus:

$$E[X_i = k \mid j \text{ follows } i] = E[X_i] + \frac{\text{Var}[X_i]}{E[X_i]} > E[X_i]$$

Since $E[X_i] = E[X_j]$ is the same for all individuals in the social network, the expected number of followers of j is less than for i (given they are followed by j).

6   Nathan O. Hodas, Farshad Kooti and Kristina Lerman, 'Friendship paradox redux: your friends are more interesting than you', in *Proceedings of the Seventh International AAAI Conference on Weblogs and Social Media*, 2013.

7   It is now published here. Michaela Norrman and Lina Hahlin, 'Hur tänker Instagram? En statistisk analys av två Instagramflöden' [How does Instagram think? A statistical analysis of two Instagram accounts] (undergraduate dissertation), Mathematics department, University of Uppsala, 2019; retrieved from <http://urn.kb.se/resolve?urn=urn:nbn:se:uu:diva-388141>.

8   Amanda Törner, 'Anitha Schulman: "Instagram går mot en beklaglig framtid"' [Instagram is heading towards an unfortunate future], Dagens Media, 5 March 2018; at <https://www.dagensmedia.se/medier/anitha-schulman-instagram-gar-mot-en-beklaglig-framtid-6902124>. (Anitha Schulman's married name is Clemence.)

9   Kelley Cotter, 'Playing the visibility game: how digital influencers and algorithms negotiate influence on Instagram', *New Media & Society* 21(4) (April 2019): 895–913.

10 Lawrence Page, 'Method for node ranking in a linked database', US Patent 6,285,999 B1, issued 4 September 2001; at <https://patentimages.storage.googleapis.com/37/a9/18/d7c46ea42c4b05/US6285999.pdf>.

6 THE MARKET EQUATION

1   See, for example, Jean-Philippe Bouchaud, 'Power laws in economics and finance: some ideas from physics', *Quantitative Finance* 1(1) (September 2000): 105–12; Rosario N. Mantegna and H. Eugene Stanley, 'Turbulence and financial markets', *Nature* 383(6601) (October 1996): 587.

2   Note that $\sqrt{n} = n^{1/2}$. So $n^{2/3}$ is bigger than $n^{1/2}$, provided $n > 1$.

3   Nassim Nicholas Taleb, *Fooled by Randomness: The Hidden Role of Chance in Life and in the Markets* (London: Random House, 2005);

Nassim Nicholas Taleb, *The Black Swan: The Impact of the Highly Improbable* (London: Allen Lane, 2007); Robert J. Shiller, *Irrational Exuberance*, revised and expanded third edition (Princeton, NJ: Princeton University Press, 2015).

4 David M. Cutler, James M. Poterba and Lawrence H. Summers, 'What moves stock prices?', NBER Working Paper No. 2538, National Bureau of Economic Research, March 1988.

5 Paul C. Tetlock, 'Giving content to investor sentiment: the role of media in the stock market', *Journal of Finance* 62(3) (2007): 1139–68.

6 Werner Antweiler and Murray Z. Frank, 'Is all that talk just noise? The information content of Internet stock message boards', *Journal of Finance* 59(3) (2004): 1259–94.

7 John Detrixhe, 'Don't kid yourself – nobody knows what really triggered the market meltdown', *Quartz*, 13 February 2018; at <https://qz.com/1205782/nobody-really-knows-why-stock-markets-went-haywire-last-week/>.

8 He published his results in the following article: Greg Laughlin, 'Insights into high frequency trading from the Virtu initial public offering', paper published online 2015; at <https://online.wsj.com/public/resources/documents/VirtuOverview.pdf>; see also Bradley Hope, 'Virtu's losing day was 1-in-1,238: odds say it shouldn't have happened at all', *Wall Street Journal*, 13 November 2014; at <https://blogs.wsj.com/moneybeat/2014/11/13/virtus-losing-day-was-1-in-1238-odds-says-it-shouldnt-have-happened-at-all/>.

9 Sam Mamudi, 'Virtu touting near-perfect record of profits backfired, CEO says', *Bloomberg News*, 4 June 2014; at <http://www.bloomberg.com/news/2014-06-04/virtu-touting-near-perfect-record-of-profits-backfired-ceo-says.html>.

10 444,000/0.0027 = 164,444,444.

11 Name changed to protect identity.

12 Paul Krugman, 'Three Expensive Milliseconds', *New York Times*, 13 April 2014; at <https://www.nytimes.com/2014/04/14/opinion/krugman-three-expensive-milliseconds.html>.

7 THE ADVERTISING EQUATION

1 For more details, see <https://medium.com/me/stats/post/2904fa0571bd>.

2 Snapchat Marketing, 'The 17 types of Snapchat users', 7 June 2016; at <http://www.snapchatmarketing.co/types-of-snapchat-users/>.

3 Noah A. Rosenberg, Jonathan K. Pritchard, James L. Weber, Howard M. Cann, Kenneth K. Kidd, Lev A. Zhivotovsky and Marcus W. Feldman, 'Genetic structure of human populations', *Science* 298(5602) (December 2002): 2381–5.

4 Shepherd Laughlin, 'Gen Z goes beyond gender binaries in new Innovation Group data', *J. Walter Thompson Intelligence*, 11 March 2016; at <https://www.jwtintelligence.com/2016/03/gen-z-goes-beyond-gender-binaries-in-new-innovation-group-data/>.

5 See, for example, Ronald Inglehart and Wayne E. Baker, 'Modernization, cultural change, and the persistence of traditional values', *American Sociological Review* 65(1) (February 2000): 19–51.

6 Ronald Inglehart and Christian Welzel, *Modernization, Cultural Change, and Democracy: The Human Development Sequence* (Cambridge: Cambridge University Press, 2005).

7 Michele Dillon, 'Asynchrony in attitudes toward abortion and gay rights: the challenge to values alignment', *Journal for the Scientific Study of Religion* 53(1) (March 2014): 1–16.

8 Anja Lambrecht and Catherine E. Tucker, 'On storks and babies: correlation, causality and field experiments', *GfK Marketing Intelligence Review* 8(2) (November 2016): 24–9.

9 David Sumpter, *Outnumbered: From Facebook and Google to Fake News and Filter-Bubbles – The Algorithms that Control Our Lives* (London: Bloomsbury Publishing, 2018).

10 Cathy O'Neil, *Weapons of Math Destruction: How Big Data Increases Inequality and Threatens Democracy* (New York: Crown Publishing Group, 2016).

11 Carole Cadwalladr, 'Google, democracy and the truth about internet search', *The Guardian*, 4 December 2016; at <https://www.theguardian.com/technology/2016/dec/04/google-democracy-truth-internet-search-facebook>.

12 Aylin Caliskan, Joanna J. Bryson and Arvind Narayanan, 'Semantics derived automatically from language corpora contain human-like biases', *Science* 356(6334) (2017): 183–6.

13 Julia Angwin, Ariana Tobin and Madeleine Varner, 'Facebook (still) letting housing advertisers exclude users by race', *ProPublica*, 21 November 2017; at <https://www.propublica.org/article/facebook-advertising-discrimination-housing-race-sex-national-origin>.

14 Anja Lambrecht, Catherine Tucker and Caroline Wiertz, 'Advertising to early trend propagators: evidence from Twitter', *Marketing Science* 37(2) (March 2018): 177–99.

8 THE REWARD EQUATION

1   Herbert Robbins and Sutton Monro, 'A stochastic approximation method', *Annals of Mathematical Statistics* 22(3) (September 1951): 400–407.

2   The full calculation is as follows:

$$Q_{10} = 0.9 \cdot 1.000 + 0.1 \cdot 0 = 0.900$$
$$Q_{11} = 0.9 \cdot 0.900 + 0.1 \cdot 1 = 0.910$$
$$Q_{12} = 0.9 \cdot 0.910 + 0.1 \cdot 1 = 0.919$$
$$Q_{13} = 0.9 \cdot 0.919 + 0.1 \cdot 0 = 0.827$$
$$Q_{14} = 0.9 \cdot 0.827 + 0.1 \cdot 0 = 0.744$$
$$Q_{15} = 0.9 \cdot 0.744 + 0.1 \cdot 1 = 0.770$$
$$Q_{16} = 0.9 \cdot 0.770 + 0.1 \cdot 0 = 0.693$$
$$Q_{17} = 0.9 \cdot 0.693 + 0.1 \cdot 1 = 0.724$$

3   Wolfram Schultz, 'Predictive reward signal of dopamine neurons', *Journal of Neurophysiology* 80(1) (July 1998): 1–27.

4   For a more thorough review of the coupling between dopamine neurons and mathematical models, see Yael Niv, 'Reinforcement learning in the brain', *Journal of Mathematical Psychology* 53(3) (June 2009): 139–54.

5   Andrew K. Przybylski, C. Scott Rigby and Richard M. Ryan, 'A motivational model of video game engagement', *Review of General Psychology* 14(2) (June 2010): 154–66.

6   Dr Emily Collins speaking about digital games and mindfulness apps, EurekAlert!, University of Bath; at <https://www.eurekalert.org/multimedia/pub/207686.php>.

7   Rudolf Emil Kálmán, 'A new approach to linear filtering and prediction problems', *Journal of Basic Engineering* 82(1) (1960): 35–45.

8   François Auger, Mickael Hilairet, Josep M. Guerrero, Eric Monmasson, Teresa Orlowska-Kowalska and Seiichiro Katsura, 'Industrial applications of the Kálmán filter: a review', *IEEE Transactions on Industrial Electronics* 60(12) (December 2013): 5458–71.

9   Irmgard Flügge-Lotz, C. F. Taylor and H. E. Lindberg, *Investigation of a Nonlinear Control System*, Report 1391 for the National Advisory Committee for Aeronautics (Washington DC: US Government Printing Office, 1958).

10 One of the most influential researchers in the area and the person who formalized the model is Jean-Louis Deneubourg. An historical starting

point is the paper by Simon Goss, Serge Aron, Jean-Louis Deneubourg and Jacques Marie Pasteels, 'Self-organized shortcuts in the Argentine ant', *Naturwissenschaften* 76(12) (1989): 579–81.

11 We could also write down an equation for the other tracking variable. It would look like this, and would track the rewards for the alternative option:

$$Q'_{t+1} = (1 - \alpha)Q'_t + \alpha \left(\frac{(Q'_t + \beta)^2}{(Q_t + \beta)^2 + (Q'_t + \beta)^2} \right) R'_t$$

12 See, for example, Malcolm Gladwell, *The Tipping Point: How Little Things Can Make a Big Difference* (Boston, MA: Little, Brown, 2000); and Philip Ball, *Critical Mass: How One Thing Leads to Another* (London: Heinemann, 2004).

13 Audrey Dussutour, Stamatios C. Nicolis, Grace Shephard, Madeleine Beekman and David J. T. Sumpter, 'The role of multiple pheromones in food recruitment by ants', *Journal of Experimental Biology* 212(15) (August 2009): 2337–48.

14 Tristan Harris, 'How technology is hijacking your mind – from a magician and Google design ethicist', Medium, 18 May 2016; at <https://medium.com/thrive-global/how-technology-hijacks-peoples-minds-from-a-magician-and-google-s-design-ethicist-56d62ef5edf3>.

15 John R. Krebs, Alejandro Kacelnik and Peter D. Taylor, 'Test of optimal sampling by foraging great tits', *Nature* 275(5675) (September 1978): 27–31.

16 Brian D. Loader, Ariadne Vromen and Michael A. Xenos, 'The networked young citizen: social media, political participation and civic engagement', *Information, Communication & Society* 17(2) (January 2014): 143–50.

17 Anna Dornhaus has studied this extensively. One example is D. Charbonneau, N. Hillis and Anna Dornhaus, '"Lazy" in nature: ant colony time budgets show high "inactivity" in the field as well as in the lab', *Insectes Sociaux* 62(1) (February 2014): 31–5.

9 THE LEARNING EQUATION

1 Paul Covington, Jay Adams and Emre Sargin, 'Deep neural networks for YouTube recommendations', conference paper, *Proceedings of the 10th ACM Conference on Recommender Systems*, September 2016, pp. 191–8.

2   Celie O'Neil-Hart and Howard Blumenstein, 'The latest video trends: where your audience is watching', Google, *Video, Consumer Insights*; at<https://www.thinkwithgoogle.com/consumer-insights/video-trends-where-audience-watching/>.

3   Chris Stokel-Walker, 'Algorithms won't fix what's wrong with You-Tube', *New York Times*, 14 June 2019; at <https://www.nytimes.com/2019/06/14/opinion/youtube-algorithm.html>.

4   K. G. Orphanides, 'Children's YouTube is still churning out blood, sui-cide and cannibalism', *Wired*, 23 March 2018; at <https://www.wired.co.uk/article/youtube-for-kids-videos-problems-algorithm-recommend>.

5   Max Fisher and Amanda Taub, 'On YouTube's digital playground, an open gate for pedophiles', *New York Times*, 3 June 2019; at <https://www.nytimes.com/2019/06/03/world/americas/youtube-pedophiles.html?module=inline>.

6   David Silver, Aja Huang, Chris J. Maddison, Arthur Guez, Laurent Sifre, George van den Driessche, Julian Schrittwieser et al., 'Mastering the game of Go with deep neural networks and tree search', *Nature* 529 (7587) (January 2016): 484–9.

7   The other, called Softmax, is very similar to Equation 1 but can be easier to work with in some situations. In most cases Softmax and Equation 1 can be used interchangeably.

8   Volodymyr Mnih, Koray Kavukcuoglu, David Silver, Andrei A. Rusu, Joel Veness, Marc G. Bellemare, Alex Graves et al., 'Human-level con-trol through deep reinforcement learning', *Nature* 518(7540) (February 2015): 529–33.

9   Tomáš Mikolov, Martin Karafiát, Lukáš Burget, Jan Černocký and San-jeev Khudanpur, 'Recurrent neural network based language model', conference paper, *Interspeech 2010*, Eleventh Annual Conference of the International Speech Communication Association, Japan, September 2010.

10 THE UNIVERSAL EQUATION

1   Thomas J. Misa and Philip L. Frana, 'An interview with Edsger W. Dijkstra', *Communications of the ACM* 53(8) (2010): 41–7.

2   See the excellent textbook by Thomas H. Cormen, Charles E. Leiser-son, Ronald L. Rivest and Clifford Stein, *Introduction to Algorithms*, third edition (Cambridge, MA: MIT Press, 2009).

3   Po-Shen Loh, *The Most Beautiful Equation in Math*, video, Carnegie Mellon University, March 2016; at <https://www.youtube.com/watch?v=IUTGFQpKaPU>.

4   The triangle is in Euclidean geometry.

5   Ben Rogers, *A. J. Ayer: A Life* (London: Chatto and Windus, 1999).

6   Philippa Foot, 'The problem of abortion and the doctrine of double effect', *Oxford Review* 5 (1967): 5–15.

7   Henrik Ahlenius and Torbjörn Tännsjö, 'Chinese and Westerners respond differently to the trolley dilemmas', *Journal of Cognition and Culture* 12(3–4) (January 2012): 195–201.

8   John Mikhail, 'Universal moral grammar: theory, evidence and the future', *Trends in Cognitive Sciences* 11(4) (April 2007): 143–52.

9   Judith Jarvis Thomson, 'Killing, letting die, and the trolley problem', *The Monist* 59(2) (1976): 204–17. The text describing the trolley problem used in the main text is taken from this article.

10 For more on the philosophical aspects of the trolley problem and the idea of moral intuition, see Laura D'Olimpio's article 'The trolley dilemma: would you kill one person to save five?', *The Conversation*, 3 June 2016; at <https://theconversation.com/the-trolley-dilemma-would-you-kill-one-person-to-save-five-57111>.

11 In Chapters 3 and 5, in order to allow the history of TEN to be told as it happened, I didn't fully explain why Richard Price's argument about miracles is wrong. The scientific evidence against miracles, such as the Resurrection, comes from an underlying understanding of biology and not just from the fact that no one has achieved such a feat since the reports of Jesus' return to life 2,000 years ago. We should view Price's contribution as a way of tightening up our thinking about evidence, rather than a proof that the Resurrection could have happened. His argument provides a genuine and important everyday lesson: the fact we haven't seen a rare event in the past should not be used to rule it out from happening in the future. It doesn't stand up to a scientific analysis combining data and model. The Resurrection can only be explained by Jesus not being dead or his death being misreported.

12 Viktoria Spaiser, Peter Hedström, Shyam Ranganathan, Kim Jansson, Monica K. Nordvik and David J. T. Sumpter, 'Identifying complex dynamics in social systems: a new methodological approach applied to study school segregation', *Sociological Methods & Research* 47(2) (March 2018): 103–35.

13 Anne's calculation is part of this report: 'UK's carbon footprint 1997–2016: annual carbon dioxide emissions relating to UK consumption', 13 December 2012, Department for Environment, Food & Rural Affairs; at <https://www.gov.uk/government/statistics/uks-carbon-footprint>.

Acknowledgements

This book started properly for me with a challenge from Helen Conford. She told me to stop writing for other people and write what I really wanted to say. I told her that I am not a very interesting person and she told me that she would be the judge of that. So, I did what she said.

I'm still not convinced that I am particularly interesting, but I do know that she and later Casiana Ionita have helped me take what I really wanted to say and make it interesting. Here the credit goes primarily to Casiana, whose simultaneously sensitive and brutal editing has made this book what it is. Thank you.

I have learnt so much from my super-agent Chris Wellbelove about writing and structuring ideas. He has even found mistakes in the maths. It's difficult to explain, but when I write I often hear Casiana, Chris and Helen bickering in my head. Thank you for those discussions that you never had.

Thank you to Jane Robertson for her careful copy-editing, Boris Granovskiy for checking the maths one extra time and to Ruth Pietroni and her team at Penguin for putting it all together.

A big thank-you to Rolf Larsson for his extra careful reading of the book and for finding one 'serious error' and several smaller ones. Thanks also to Oliver Johnson for his careful feedback and for his suggestion of Figure 2.

I write best when I am surrounded by activity and life. So, thank you to Hammarby Football, Uppsala Mathematics Department, my daughter Elise and son Henry and friends, especially the Pellings, who have provided that life during the last year.

Thanks to my dad for introducing me to A. J. Ayer. To my mum, I'm sorry you got 'cut', but everything I wrote about you remains true: you are an inspiration to everyone around you. Thanks also to both of you for the extensive and thoughtful comments.

Most of all, I want to thank Lovisa. Often what I really want to say is about our life, our discussions, our agreements and our arguments. I hope a few of these have found their way into this book.

ALLEN LANE
an imprint of
PENGUIN BOOKS

Also Published

Lisa Miller, *The Awakened Brain: The Psychology of Spirituality and Our Search for Meaning*

Michael Pye, *Antwerp: The Glory Years*

Christopher Clark, *Prisoners of Time: Prussians, Germans and Other Humans*

Rupa Marya and Raj Patel, *Inflamed: Deep Medicine and the Anatomy of Injustice*

Richard Zenith, *Pessoa: An Experimental Life*

Michael Pollan, *This Is Your Mind On Plants: Opium—Caffeine—Mescaline*

Amartya Sen, *Home in the World: A Memoir*

Jan-Werner Müller, *Democracy Rules*

Robin DiAngelo, *Nice Racism: How Progressive White People Perpetuate Racial Harm*

Rosemary Hill, *Time's Witness: History in the Age of Romanticism*

Lawrence Wright, *The Plague Year: America in the Time of Covid*

Adrian Wooldridge, *The Aristocracy of Talent: How Meritocracy Made the Modern World*

Julian Hoppit, *The Dreadful Monster and its Poor Relations: Taxing, Spending and the United Kingdom, 1707-2021*

Jordan Ellenberg, *Shape: The Hidden Geometry of Absolutely Everything*

Duncan Campbell-Smith, *Crossing Continents: A History of Standard Chartered Bank*

Jemma Wadham, *Ice Rivers*

Niall Ferguson, *Doom: The Politics of Catastrophe*

Michael Lewis, *The Premonition: A Pandemic Story*

Chiara Marletto, *The Science of Can and Can't: A Physicist's Journey Through the Land of Counterfactuals*

Suzanne Simard, *Finding the Mother Tree: Uncovering the Wisdom and Intelligence of the Forest*

Giles Fraser, *Chosen: Lost and Found between Christianity and Judaism*

Malcolm Gladwell, *The Bomber Mafia: A Story Set in War*

Kate Darling, *The New Breed: How to Think About Robots*

Serhii Plokhy, *Nuclear Folly: A New History of the Cuban Missile Crisis*

Sean McMeekin, *Stalin's War*

Michio Kaku, *The God Equation: The Quest for a Theory of Everything*

Michael Barber, *Accomplishment: How to Achieve Ambitious and Challenging Things*

Charles Townshend, *The Partition: Ireland Divided, 1885-1925*

Hanif Abdurraqib, *A Little Devil in America: In Priase of Black Performance*

Carlo Rovelli, *Helgoland*

Herman Pontzer, *Burn: The Misunderstood Science of Metabolism*

Jordan B. Peterson, *Beyond Order: 12 More Rules for Life*

Bill Gates, *How to Avoid a Climate Disaster: The Solutions We Have and the Breakthroughs We Need*

Kehinde Andrews, *The New Age of Empire: How Racism and Colonialism Still Rule the World*

Veronica O'Keane, *The Rag and Bone Shop: How We Make Memories and Memories Make Us*

Robert Tombs, *This Sovereign Isle: Britain In and Out of Europe*

Mariana Mazzucato, *Mission Economy: A Moonshot Guide to Changing Capitalism*

Frank Wilczek, *Fundamentals: Ten Keys to Reality*

Milo Beckman, *Math Without Numbers*

John Sellars, *The Fourfold Remedy: Epicurus and the Art of Happiness*

T. G. Otte, *Statesman of Europe: A Life of Sir Edward Grey*

Alex Kerr, *Finding the Heart Sutra: Guided by a Magician, an Art Collector and Buddhist Sages from Tibet to Japan*

Edwin Gale, *The Species That Changed Itself: How Prosperity Reshaped Humanity*

Simon Baron-Cohen, *The Pattern Seekers: A New Theory of Human Invention*

Christopher Harding, *The Japanese: A History of Twenty Lives*

Carlo Rovelli, *There Are Places in the World Where Rules Are Less Important Than Kindness*

Ritchie Robertson, *The Enlightenment: The Pursuit of Happiness 1680-1790*

Ivan Krastev, *Is It Tomorrow Yet?: Paradoxes of the Pandemic*

Tim Harper, *Underground Asia: Global Revolutionaries and the Assault on Empire*

John Gray, *Feline Philosophy: Cats and the Meaning of Life*

Priya Satia, *Time's Monster: History, Conscience and Britain's Empire*

Fareed Zakaria, *Ten Lessons for a Post-Pandemic World*

David Sumpter, *The Ten Equations that Rule the World: And How You Can Use Them Too*

Richard J. Evans, *The Hitler Conspiracies: The Third Reich and the Paranoid Imagination*

Fernando Cervantes, *Conquistadores*

John Darwin, *Unlocking the World: Port Cities and Globalization in the Age of Steam, 1830-1930*

Michael Strevens, *The Knowledge Machine: How an Unreasonable Idea Created Modern Science*

Owen Jones, *This Land: The Story of a Movement*

Seb Falk, *The Light Ages: A Medieval Journey of Discovery*

Daniel Yergin, *The New Map: Energy, Climate, and the Clash of Nations*

Michael J. Sandel, *The Tyranny of Merit: What's Become of the Common Good?*

Joseph Henrich, *The Weirdest People in the World: How the West Became Psychologically Peculiar and Particularly Prosperous*

Leonard Mlodinow, *Stephen Hawking: A Memoir of Friendship and Physics*

David Goodhart, *Head Hand Heart: The Struggle for Dignity and Status in the 21st Century*

Claudia Rankine, *Just Us: An American Conversation*

James Rebanks, *English Pastoral: An Inheritance*

Robin Lane Fox, *The Invention of Medicine: From Homer to Hippocrates*

Daniel Lieberman, *Exercised: The Science of Physical Activity, Rest and Health*

Sudhir Hazareesingh, *Black Spartacus: The Epic Life of Touissaint Louverture*

Judith Herrin, *Ravenna: Capital of Empire, Crucible of Europe*

Samantha Cristoforetti, *Diary of an Apprentice Astronaut*

Neil Price, *The Children of Ash and Elm: A History of the Vikings*

George Dyson, *Analogia: The Entangled Destinies of Nature, Human Beings and Machines*

Wolfram Eilenberger, *Time of the Magicians: The Invention of Modern Thought, 1919-1929*

Kate Manne, *Entitled: How Male Privilege Hurts Women*

Christopher de Hamel, *The Book in the Cathedral: The Last Relic of Thomas Becket*

Isabel Wilkerson, *Caste: The International Bestseller*

Bradley Garrett, *Bunker: Building for the End Times*

Katie Mack, *The End of Everything: (Astrophysically Speaking)*

Jonathan C. Slaght, *Owls of the Eastern Ice: The Quest to Find and Save the World's Largest Owl*

Carl T. Bergstrom and Jevin D. West, *Calling Bullshit: The Art of Scepticism in a Data-Driven World*

Paul Collier and John Kay, *Greed Is Dead: Politics After Individualism*

Anne Applebaum, *Twilight of Democracy: The Failure of Politics and the Parting of Friends*

Sarah Stewart Johnson, *The Sirens of Mars: Searching for Life on Another World*

Martyn Rady, *The Habsburgs: The Rise and Fall of a World Power*

John Gooch, *Mussolini's War: Fascist Italy from Triumph to Collapse, 1935-1943*

Roger Scruton, *Wagner's Parsifal: The Music of Redemption*

Roberto Calasso, *The Celestial Hunter*

Benjamin R. Teitelbaum, *War for Eternity: The Return of Traditionalism and the Rise of the Populist Right*

Laurence C. Smith, *Rivers of Power: How a Natural Force Raised Kingdoms, Destroyed Civilizations, and Shapes Our World*

Sharon Moalem, *The Better Half: On the Genetic Superiority of Women*

Augustine Sedgwick, *Coffeeland: A History*

Daniel Todman, *Britain's War: A New World, 1942-1947*

Anatol Lieven, *Climate Change and the Nation State: The Realist Case*

Blake Gopnik, *Warhol: A Life as Art*

Malena and Beata Ernman, Svante and Greta Thunberg, *Our House is on Fire: Scenes of a Family and a Planet in Crisis*

Paolo Zellini, *The Mathematics of the Gods and the Algorithms of Men: A Cultural History*

Bari Weiss, *How to Fight Anti-Semitism*

Lucy Jones, *Losing Eden: Why Our Minds Need the Wild*

Brian Greene, *Until the End of Time: Mind, Matter, and Our Search for Meaning in an Evolving Universe*

Anastasia Nesvetailova and Ronen Palan, *Sabotage: The Business of Finance*

Albert Costa, *The Bilingual Brain: And What It Tells Us about the Science of Language*

Stanislas Dehaene, *How We Learn: The New Science of Education and the Brain*

Daniel Susskind, *A World Without Work: Technology, Automation and How We Should Respond*

John Tierney and Roy F. Baumeister, *The Power of Bad: And How to Overcome It*

Greta Thunberg, *No One Is Too Small to Make a Difference: Illustrated Edition*

Glenn Simpson and Peter Fritsch, *Crime in Progress: The Secret History of the Trump-Russia Investigation*

Abhijit V. Banerjee and Esther Duflo, *Good Economics for Hard Times: Better Answers to Our Biggest Problems*

Gaia Vince, *Transcendence: How Humans Evolved through Fire, Language, Beauty and Time*

Roderick Floud, *An Economic History of the English Garden*

Rana Foroohar, *Don't Be Evil: The Case Against Big Tech*

Ivan Krastev and Stephen Holmes, *The Light that Failed: A Reckoning*

Andrew Roberts, *Leadership in War: Lessons from Those Who Made History*

Alexander Watson, *The Fortress: The Great Siege of Przemysl*

Stuart Russell, *Human Compatible: AI and the Problem of Control*

Serhii Plokhy, *Forgotten Bastards of the Eastern Front: An Untold Story of World War II*

Dominic Sandbrook, *Who Dares Wins: Britain, 1979-1982*

Charles Moore, *Margaret Thatcher: The Authorized Biography, Volume Three: Herself Alone*

Thomas Penn, *The Brothers York: An English Tragedy*

David Abulafia, *The Boundless Sea: A Human History of the Oceans*

Anthony Aguirre, *Cosmological Koans: A Journey to the Heart of Physics*

Orlando Figes, *The Europeans: Three Lives and the Making of a Cosmopolitan Culture*

Naomi Klein, *On Fire: The Burning Case for a Green New Deal*

Anne Boyer, *The Undying: A Meditation on Modern Illness*

Benjamin Moser, *Sontag: Her Life*

Daniel Markovits, *The Meritocracy Trap*

Malcolm Gladwell, *Talking to Strangers: What We Should Know about the People We Don't Know*

Peter Hennessy, *Winds of Change: Britain in the Early Sixties*

John Sellars, *Lessons in Stoicism: What Ancient Philosophers Teach Us about How to Live*

Brendan Simms, *Hitler: Only the World Was Enough*

Hassan Damluji, *The Responsible Globalist: What Citizens of the World Can Learn from Nationalism*

Peter Gatrell, *The Unsettling of Europe: The Great Migration, 1945 to the Present*

Justin Marozzi, *Islamic Empires: Fifteen Cities that Define a Civilization*

Bruce Hood, *Possessed: Why We Want More Than We Need*

Susan Neiman, *Learning from the Germans: Confronting Race and the Memory of Evil*

Donald D. Hoffman, *The Case Against Reality: How Evolution Hid the Truth from Our Eyes*

Frank Close, *Trinity: The Treachery and Pursuit of the Most Dangerous Spy in History*

Richard M. Eaton, *India in the Persianate Age: 1000-1765*

Janet L. Nelson, *King and Emperor: A New Life of Charlemagne*

Philip Mansel, *King of the World: The Life of Louis XIV*

Donald Sassoon, *The Anxious Triumph: A Global History of Capitalism, 1860-1914*

Elliot Ackerman, *Places and Names: On War, Revolution and Returning*

Jonathan Aldred, *Licence to be Bad: How Economics Corrupted Us*

Johny Pitts, *Afropean: Notes from Black Europe*

Walt Odets, *Out of the Shadows: Reimagining Gay Men's Lives*

James Lovelock, *Novacene: The Coming Age of Hyperintelligence*

Mark B. Smith, *The Russia Anxiety: And How History Can Resolve It*

Stella Tillyard, *George IV: King in Waiting*

Jonathan Rée, *Witcraft: The Invention of Philosophy in English*

Jared Diamond, *Upheaval: How Nations Cope with Crisis and Change*

Emma Dabiri, *Don't Touch My Hair*

Srecko Horvat, *Poetry from the Future: Why a Global Liberation Movement Is Our Civilisation's Last Chance*